The Bionarrative

The story of life and hope for the future

The Bionarrative

The story of life and hope for the future

Stephen Boyden

Australian
National
University

PRESS

ANU PRESS

Published by ANU Press
The Australian National University
Acton ACT 2601, Australia
Email: anupress@anu.edu.au
This title is also available online at press.anu.edu.au

National Library of Australia Cataloguing-in-Publication entry

Creator: Boyden, Stephen, author.

Title: The bionarrative : the story of life and hope for the future
                / Stephen Boyden.

ISBN: 9781760460501 (paperback) 9781760460518 (ebook)

Subjects: Human ecology.
                Human evolution.
                Nature--Effect of human beings on.
                Human beings--Effect of environment on.

Dewey Number: 304.2

Cover design and layout by ANU Press.
Cover photographs adapted from: 'LA landscape' by Doc Searls flic.kr/p/vUvCxY
and 'Rain drops' by Alex Block flic.kr/p/a2NgMe.

Contents

Acknowledgements

First, I would like to express my heartfelt thanks to the students and academic colleagues who shared the experience of developing the discipline and teaching of human ecology and biohistory at The Australian National University (ANU), and who contributed so much to my thinking and enjoyment of life. I am also indebted to my friends in the Nature and Society Forum (now the Frank Fenner Foundation) who, since my retirement at the end of 1990, have been so much a part of my life and whose enthusiasm and creativity have been a constant source of inspiration.

I would also like to mention two individuals who made it all possible: first, the late Leonard Huxley, vice-chancellor of ANU, whose unconventional decision in 1965 allowed me to make a radical change in my field of work at the university; and second, my friend the late Frank Fenner, director of the John Curtin School of Medical Research and later of the Centre for Resource and Environmental Studies, whose unswerving support over many years enabled us to develop human ecology and biohistory as legitimate areas of study in the university.

I am grateful to the ANU Publication Subsidy Committee for the provision of funding in support of this book.

Preface

This volume is an outcome of work carried out at The Australian National University, after I turned my attention to the biology of civilisation some 50 years ago. My colleagues and I refer to our approach as 'biohistory', which we define as the study of human situations, past and present, against the background of the history of life on Earth.

We see biohistory as much more than just an interesting academic exercise. It has great practical meaning for society as a whole as well as for individuals and families.

This book is intended for the general reader who is interested in the human place in nature and the future of civilisation. It is not an academic treatise. The central theme is the story of life on Earth, and of humans and their civilisation as part of that story. I refer to this story as the 'bionarrative'. The final chapter discusses options for the future of humankind.

In telling this story, the challenge is to decide what to put in and what to leave out. To do the job thoroughly would take up many volumes. I have selected for inclusion aspects that seem to me either especially interesting or relevant for understanding the human situation on Earth today.

Although biohistory is not yet recognised as an academic discipline, a number of authors have emerged over recent decades who could well be described as biohistorians. They have told this story, or parts of it, in their own ways, emphasising different facts and principles. This book is my version of the bionarrative.

I must warn readers that I make free use of the prefix 'bio-'. Some readers may find this irritating. I find it preferable, however, to say 'biounderstanding', rather than repeating again and again 'understanding the processes of life and the human place in nature'; or 'biosensitive', rather than 'sensitive to, in tune with, and respectful of the processes of life'.

I have kept referencing to a minimum. References are given for all the quotations and I include a few other references that I feel might be particularly useful to the reader. The facts and figures presented about human health, energy and resource use, and environmental change are readily verifiable on the internet. The literature in environmental philosophy and environmental history is now so vast that it would not make sense to try to cover this in the bibliography.

Some passages of the text have already appeared in my previous publications.

Stephen Boyden
13 July 2016

1

The bionarrative, biohistory and biounderstanding

Introduction

This book presents a very brief version of the story of life on Earth. The recent emergence of humankind and of human civilisation is an integral and vitally important part of this story. It is a story of overarching significance for every one of us and for society as a whole; yet it is known and understood by only a small section of the human community. I believe that, were it to be embraced by the dominant cultures across the world, the prospects for the future of humankind would be greatly enhanced. I refer to this story as the bionarrative.

The bionarrative, as presented here, is based on a conceptual approach to the study of human situations known as biohistory. Biohistory is the study of human situations, past and present, against the background of the history of life on Earth. It covers the basic principles of evolution, ecology, inheritance, and health and disease, and it pays special attention to the evolutionary background of our species; it recognises the immense ecological importance of the emergence in evolution of humankind's most distinctive biological attribute, the capacity for language and culture.[1]

The human capacity for culture, and therefore culture itself, are products of biological evolution. Culture is created and stored in human brains, and it is entirely dependent for its existence on the processes of life within the human body and in the ecosystems that support us. Through its influence on human behaviour it has big impacts on other forms of life. Human culture is thus a biological phenomenon that is intimately connected with other parts of the living system.[2] Yet academia has separated culture off from the rest of the living world. The humanities and the life sciences are studied and taught by different groups of people who have little to do with one another.

The central theme of biohistory, then, is life — not only in the evolutionary past, but also as the source and mainstay of everything that goes on in human society today. Civilisation and all its institutions, economic arrangements, technologies and works of art are manifestations of life, and all are totally dependent on life processes.

I refer to the understanding of the bionarrative as biounderstanding. I am among those who believe that shared biounderstanding across the global community is an essential prerequisite for the future well-being of humankind. At present, however, biohistory is not recognised as a bona fide subject in academic circles. It does not appear in school curricula and it does not feature in university degree courses or research programs.[3]

---

1   The word 'culture' has many, rather different, meanings. Here it is used to mean the abstract products of the capacity for culture, such as learned language itself and the accumulated knowledge, assumptions, beliefs, values and technological know-how of a human population. This use of the term is consistent with the first definition of culture given in the *Collins Dictionary*: 'The total of the inherited ideas, beliefs, values and knowledge, which constitute the shared bases of social action' (*Collins Dictionary of the English Language* (1979), Collins, Sydney, Auckland and Glasgow).

2   Although, of course, most products of culture, like buildings, works of art, books and computers, are not part of life.

3   Over recent decades, a growing number of writers have emerged who could well be described as biohistorians. René Dubos comes first to mind. Others include, Jared Diamond and Tony McMichael. Biohistory, however, has yet to be developed systematically as a field of learning, and it is a long way from occupying the central place it warrants in educational programs at all levels.

Conceptual framework

Biohistory takes as its starting point the history of life on Earth. In the beginning there was no life and only the physical world existed. Then, perhaps around 4.5 billion years ago, the first living organisms came into being. Eventually, over many millions of years, there evolved an amazing array of different life forms. Among these, emerging some 200,000 years ago, was *Homo sapiens*.

Through the processes of biological evolution, the human species acquired a distinctive and extraordinarily significant biological attribute, which is unique in the animal kingdom. This is the ability to invent, memorise and communicate with a symbolic spoken language. The aptitude for language eventually led to the accumulation by human groups of shared knowledge, beliefs and attitudes. That is, it led to human culture.

As soon as human culture came into existence it began, through its influence on people's behaviour, to have impacts not only on humans themselves but also on other living systems. It evolved as a new kind of force in the biosphere, which was destined eventually to bring about profound and far-reaching changes across the whole planet.

A conceptual framework reflecting this approach is depicted in Figure 1.1. The arrows in this figure simply mark pathways of influence. They do not represent flows of matter or energy. Although this conceptual framework is based on the sequence of happenings in the history of life on Earth, it can also be applied to the here and now.

In this model, 'Humankind' is placed at the top of the scheme, supported by the 'Rest of the biosphere' at the base. The 'Rest of the biosphere' is made up of two broad categories of variables — the 'Physical world' and 'Living organisms'.

Humankind

Prevailing culture

Societal arrangements

Human population

Human activities

Human artefacts

Rest of Biosphere

Living organisms

Physical world

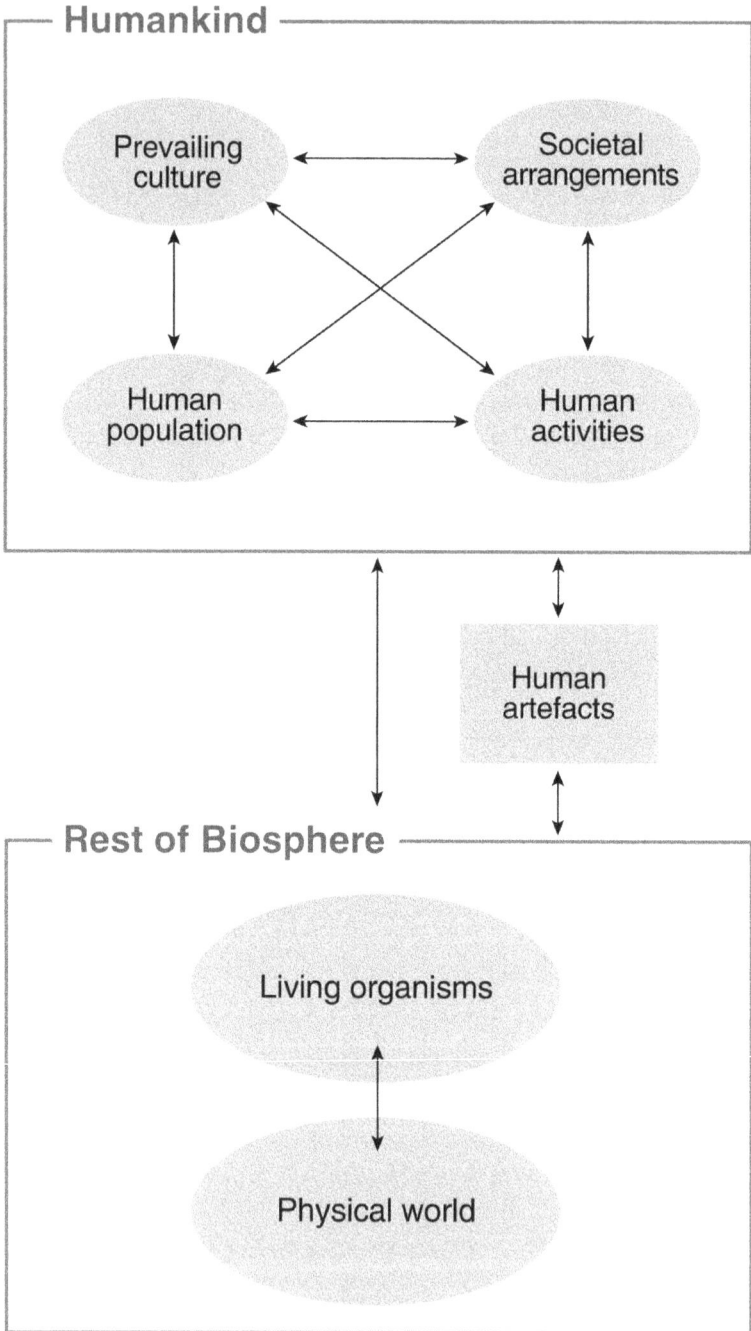

Figure 1.1 A conceptual framework

Source: Stephen Boyden

'Humankind' is separated into four sets of interacting variables as follows:

- 'Human population'
  This category of variables describes the state of the human population at any given time. It includes such variables as population size, health statistics, and geographical distribution.

- 'Human activities'
  This group of variable covers all kinds of human behaviour including, for example, farming, manufacturing, retailing, travelling, using energy of one kind or another and making war. Human activities clearly have impacts on humans themselves as well as on other living organisms. They are determined partly by the biological characteristics of humans. The prevailing culture is also a major determinant of human activities.

- 'Prevailing culture'
  This consists of the shared beliefs, knowledge (including knowledge of language, of technologies and of how things work), values and priorities of a human population. It is a major determinant of human activities and, therefore, has important indirect impacts on living systems, human and otherwise.

- 'Societal arrangements'
  This group of variables is another aspect of human culture. It includes such items as legislation, government regulations, the economic system, the institutional structure of society and educational programs. Societal arrangements are largely determined by, and to some extent determine, the characteristics of the prevailing culture.

We will have this framework in mind from Chapter 3 onwards and, especially, when considering options for the future in Chapter 7.

The framework as described above is applicable to a human population. It can also, with appropriate modifications, be applied at the level of a small group of humans or, even, at the level of the individual. In the case of the individual the 'prevailing culture' would be replaced by 'personal culture', although the prevailing culture is an important feature of an individual's personal environment.

Figure 1.1 also includes the set of variables designated 'Human artefacts', by which we mean 'things made by humans', including books, buildings, roads, machines and electronic devices, as well as clothes, utensils and works of art.

A key aspect of the biology and ecology of human ecosystems of great relevance to their sustainability is the patterns of flow of energy and materials in the biophysical system between human populations and the biosphere, between different human groups in the system, and between different components of the biosphere. The ecological sustainability of an urban settlement is, in the long term, largely a function of its pattern of inputs of materials and energy sources, and outputs of waste products. The analysis of these flows will be increasingly used by planning authorities in formulating policies for achieving ecological sustainability.

There are also flows of information within the cultural system that are of immense biological significance. These flows determine patterns of human activity, which in turn have impacts on the flows of materials and energy in the biophysical system and, in particular, between society and the rest of the biosphere.

I will now briefly draw attention to two crucially important biohistorical themes, each of which will be discussed in more detail in later chapters.

Watersheds in evolution

The evolution of life on Earth has been marked by a series of crucial watersheds, each of which changed the living world forever. Especially important among these were the development of photosynthesis, the appearance of cells with nuclei, the development of multicellularity, the beginning of sexual reproduction and the invasion of land by life forms.

The most recent crucial watershed in biological evolution was the emergence of the human capacity for language and culture. Human culture eventually developed into a new and extremely powerful force in the living world — with far-reaching consequences for life on our planet.

Cultural evolution, like biological evolution, has been marked by a series of watersheds, each of which ushered in a new and different ecological phase of human existence (Box 1.1). Although the distinctions between these four ecological phases are not always sharp, and some societies do not fit neatly into any one of the four categories, the classification is a useful one.[4]

The first of the cultural watersheds was the shared knowledge of how to make use of and, up to a point, control fire. The deliberate and regular use of fire was an important feature of ecological Phase 1 of the history of our species, the Hunter–Gatherer Phase, which lasted for around 8,000 generations.[5] During this time, *Homo sapiens* spread from Africa and, by 13,000 years ago (possibly much earlier), our species had reached all five habitable continents.

The second cultural watershed began around 12,000 years ago and led to the Early Farming Phase (about 480 generations). This was indeed a turning point in cultural evolution. Without it, the spectacular developments in human history since that time would have been impossible.

The third crucial watershed in cultural evolution was associated with the advent of urbanisation. It began around 9,000 years ago (360 generations), but really got underway about 5,000 years ago, when fully fledged cities with populations of tens of thousands were in existence in the Middle East. There were also at this time townships with populations of a few thousand in Peru. This was the beginning of ecological Phase 3 — the Early Urban Phase. For the first time in human history, very large numbers of people were separated from the natural environment and played no role in the acquisition of food. The culture and ecology of these urban dwellers were significantly different from those of hunter–gatherers or early farmers.

---

4   The emphasis here is on ecologically significant watersheds. There were also cultural watersheds affecting other aspects of human society. For example, in the sphere of the communication and storage of information, the introduction of writing and, recently, of information technology, were hugely significant watersheds.

5   For the purpose of this discussion, a generation is taken to be 25 years.

Box 1.1 The four ecological phases of human history*

	Phase 1 Hunter–Gatherer	Phase 2 Early Farming	Phase 3 Early Urban	Phase 4 Exponential
Beginning (years ago)	200,000	12,000	8,000	200
Global population** (millions)	5 (?)	50	1,000	8,000
Main causes of death	Injury Infected wounds Predators	Infected wounds Malaria Schistosomiasis	Infectious disease Malnutrition	Cardiovascular disease Cancer
Energy use — (MJ) per capita, per day	20	30	40	1,000***
Social disparities	±	+	++++	++++
'Economic growth'	-	-	+	++++
Interaction with natural environment	++++	++++	±	±
Occupational specialisation	±	+	++++	++++
Population density	±	+	++++	++++
Daily contact with strangers	±	±	++++	++++

* All figures are, of course, approximate, and the + signs are notional
** Immediately before the start of the next ecological phase
*** This figure is for a typical affluent Western country at the present time
Source: Stephen Boyden

The fourth cultural watershed consisted of the 18th-century philosophical movement referred to, misguidedly, as the Enlightenment. I say misguidedly because a more appropriate term would be Partial Enlightenment. The great weakness of the Enlightenment lay in its association with the idea that nature is out there to be conquered.

This fourth cultural watershed led to ecological Phase 4, the Exponential Phase. It has also been called the High Energy Phase or Techno-Industrial Phase, and it has recently been dubbed the Anthropocene.

Ecological Phase 4 is characterised by an exponential increase in the scale and intensity of human activities on Earth. The present pattern is not sustainable ecologically and Phase 4 is destined to come to an end very soon. Its days are numbered; business as usual will lead to the ecological collapse of civilisation.

The four phases are not mutually exclusive and all four can exist at the same time. A few hunter–gatherer societies still exist today, although most of them have been influenced by contact with people from exponential societies. Early farming societies continued to exist throughout the Early Urban Phase, providing city dwellers with food.

Adaptation and maladaptation

In biology, adaptation is defined as the process of change by which an organism or species becomes better suited to its environment. There are several different kinds of adaptation in biological systems. Particularly important are genetic, or evolutionary, adaptation and physiological adaptation.

Genetic adaptation is the kind of adaptation that has given rise to all the species of animals and plants on Earth today, including humankind. It is transgenerational and the main influence on its direction is natural selection.

Physiological adaptation consists of physiological changes in living organisms that render them better able to cope with an existing situation or threat. The heart will beat faster in a threatening situation so that muscles are provided with more oxygen and perform better if needed (e.g. in running away or fighting). Another example is the immune response that enables organisms to fight off invading microorganisms.

In humankind there is another dimension to adaptation: cultural adaptation. Cultural adaptation can be defined as cultural changes that result in humans becoming better suited to their environment. The deliberate use of fire for cooking and as a source of warmth is an early example, as is the later introduction of farming.

There is, however, another side to the picture. Not all changes in genetic material are beneficial and, in fact, the great majority of mutations are harmful. Such instances are referred to as genetic maladaptations.

Similarly, physiological responses can sometimes be harmful. Autoimmune disease is a clear example of physiological maladaptation.

So, too, with culture. As cultures have evolved they have often come to embrace not only factual information of good practical value, but also assumptions that are sheer nonsense, leading to behaviours that are equally nonsensical. That is, cultures often get things wrong. Sometimes these cultural delusions have resulted in activities that have caused unnecessary distress in humans or unnecessary damage to local ecosystems. Such cases are examples of cultural maladaptation.

There have been countless instances of cultural maladaptation in human history, and some of these will be discussed later in this book. They include religious genocide, slavery and European imperialism — all of which have been accepted as perfectly reasonable by large numbers of people for long periods of time.

Biohistory thus alerts us to the need for us to be constantly vigilant — making sure that the assumptions of our society's prevailing culture are in tune with the processes of life, and that they are not leading us to behave in ways that are against nature or against the interests of humankind.

As we will discuss in later chapters, some cultural maladaptations today are on a scale and of a kind that threaten the continued existence of human civilisation.

2

Life before humans

Planet Earth

Our planet is about 4.5 billion years old. Situated 150 million kilometres away from the Sun, it has a circumference of about 30,800 kilometres. It is almost spherical, but bulges a little at the equator and it is flattened slightly at the North and South poles.

The Earth has an inner core, which makes up about 16 per cent of its total volume, and an outer mantle. The core consists mainly of molten iron at a temperature of around 2,500°C, although there appears to be a solid part right at the centre. The mantle is about 2,900 kilometres thick and consists of relatively solid rock (Figure 2.1).

Outside the mantle there is a relatively thin layer of less dense rock called the crust. The average thickness of the crust is 35 kilometres on land and 5–6 kilometres under the oceans. The highest point on the Earth's crust, Mount Everest, is nearly 19 kilometres above the lowest point, which is in the ocean just off the coast of the Philippines.

Ninety-eight per cent of the solid matter of the Earth's crust is made up of eight elements. These, in order of abundance, are oxygen, silicon, aluminium, iron, calcium, sodium, potassium and magnesium. Oxygen accounts for 94 per cent of the crust by volume and 47 per cent by weight, and silicon accounts for 1 per cent of the crust by volume and 28 per cent by weight.

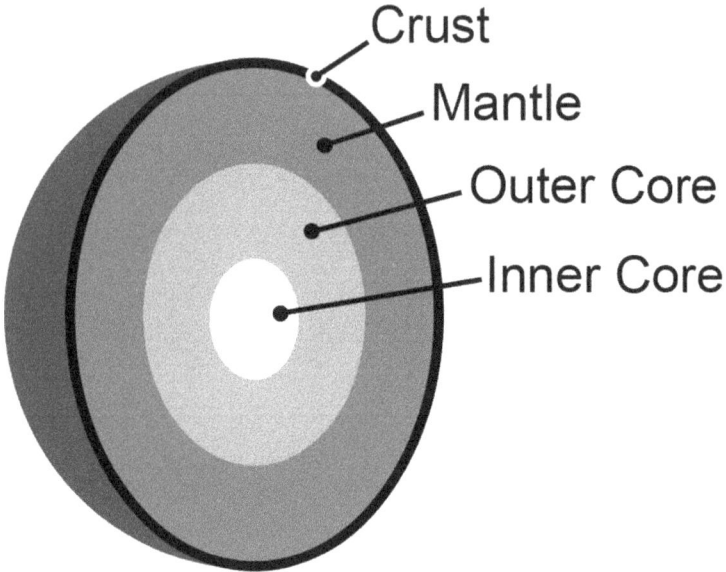

Figure 2.1 Planet Earth
Source: Stephen Boyden

There are three broad classes of rock in the Earth's crust. First, there is the original igneous rock, which was formed when the hot molten liquid of the Earth's core cooled and solidified. Basalt, obsidian and granite are examples of igneous rock. Second, there is sedimentary rock, which was formed by pressure or a chemical cementing action on rock fragments. Examples of sedimentary rock include sandstone and shale.

The third main category is metamorphic rock, in which the original structure has been altered by the action of heat, pressure or chemicals. Examples of metamorphic rock include schist, slate and marble.

As a result of physical and chemical weathering, some of the rock of the crust is broken up, forming a layer of particles of disintegrated rock of different sizes, like gravel, sand and clay.

The Earth's crust is coated with an envelope of gases — the atmosphere. Apart from water vapour, the main permanent gases in the atmosphere are nitrogen (78 per cent), oxygen (21 per cent), argon (0.93 per cent) and carbon dioxide (0.04 per cent and increasing).

Another key component of the planet's surface is water, around 97 per cent of which is in the oceans. About 2 per cent of the Earth's surface water is in the form of ice or snow in the polar regions, and about 0.5 to 1.5 per cent in the soil and in cracks between rocks. Less than 0.03 per cent is in ponds, streams, rivers and lakes, while only about 0.001 per cent is in the atmosphere.

The surface of the Earth receives constant radiation of energy from the Sun, where it is generated by nuclear fusion. This input of energy is in the form of the shorter wavelength ultraviolet rays, through visible light, to infrared. It is eventually re-radiated back into space, mainly in the form of heat.

The energy from the Sun is largely responsible for the two great circulatory systems on the Earth's surface — those of the atmosphere and the oceans. The flows in the atmosphere are caused by the unequal heating of large masses of air, and this leads to air movements that then set in motion the flows of water in the surface layers of the oceans. The patterns of flow in both the atmosphere and the oceans are also affected by the rotational movement of the planet. This is known as the Coriolis effect. The end result is that heat becomes more evenly spread over the surface of the planet.

Another important process that is driven partly by the energy from the Sun, but also by gravity, is the *water cycle*. Heat from the Sun causes water to evaporate from the surfaces of the oceans, lakes and land to form water vapour in the atmosphere. When this vapour cools, the water condenses and gravity eventually causes the droplets to fall back to Earth as rain or snow. Gravity also plays an essential role by causing much of the water that falls on land to sink down into the soil and then move into streams and rivers, from where it eventually flows back into the oceans.

Certain of the gases in the atmosphere, notably water vapour and, to a lesser extent, carbon dioxide, play a key role in keeping the temperature of the Earth's surface at levels suitable for life as we know it. The end result of this process, which is referred to as the greenhouse effect, is a world with an average temperature of around around 15°C. If these gases were not there, the energy radiated onto the Earth's surface from the Sun would re-radiate back into space, mainly in the form of heat, and the average temperature of the Earth would be −18°C.

The evolution of life

The first 4 billion years

The earliest living things on Earth are believed to have come into being around 4 billion years ago. They were single-celled organisms and they were the most complex form of life on Earth for approximately a 1 billion years. There were, and still are, two distinct groups of such microorganisms with different biochemical characteristics. They are classified as Bacteria and Archaea. The Archaea include microbes that live and multiply under extreme conditions, such as very high temperatures and very high salinity. It is not known which of these groups came into existence first.

It is believed that the main source of energy for the first single-celled organisms was energy-containing chemical compounds that had formed through the action of ultraviolet (UV) radiation and electrical discharges in storms. But the amount of energy from such sources was strictly limited, and there was certainly not enough of it to sustain life on the scale that exists today.

Single-celled organisms capable of photosynthesis — cyanobacteria — were in existence by around 2.5 billion years ago. This development represents one of the great watersheds in biological evolution. It changed the living world forever. In photosynthesis, light energy from the Sun is captured in the leaves of green plants and converted into chemical energy in the form of complex organic molecules. All animal and plant life on Earth is entirely dependent on this process. Photosynthesis involves the uptake of carbon dioxide and water from the environment and the release of free oxygen.

The emergence of photosynthesis had far-reaching evolutionary consequences. Among these was the fact that oxygen began to accumulate in the atmosphere, making it possible for life forms to evolve that relied on oxygen for their respiratory processes. Another outcome was the fact that some of the atmospheric oxygen was converted to ozone (O_3), which formed a layer in the upper part of the atmosphere. Here, it acted as a filter, absorbing much of the UV radiation from the Sun. As a result, by the time that humans appeared on Earth, and probably by 2 billion years before this, only about half of the total solar UV radiation, and a much smaller fraction of the short-wave UV-B rays, penetrated through to the Earth's surface. Had it not been for this effect, life as it exists on land today would not have been possible.

Although excessive UV radiation is damaging to living organisms, the UV rays that continue to penetrate beyond the ozone layer play a number of useful biological roles, including the promotion of the synthesis of vitamin D in human skin.

Like bacteria today, the earliest single-celled organisms did not possess nuclei. The first nucleated cells appeared about 1.5 billion years ago, and it seems that, around this time, a great evolutionary diversification began to take place among living forms, which suggests that a form of sexual reproduction was by then in existence. Previously, all reproduction had been asexual, involving the simple division of one cell into two. In sexual reproduction a new individual comes into existence through the union of two cells, the male and female *gametes*, each bringing its complement of genetic material (deoxyribonucleic acid (DNA)) from one of the parent organisms.

Around 600 to 700 million years ago, another watershed occurred in the history of life on our planet in the appearance of multicellular organisms. There is uncertainty about the timing of this evolutionary development but, by about 700 million years ago, there were flat and soft-bodied multicellular creatures in existence. They are called Ediacarans, after the Ediacara Hills in South Australia, where the first big deposits of their fossils were found.

By 500 to 600 million years ago, the Ediacarans had been replaced by very different fauna and flora, which included seaweeds, sponges, jellyfish, corals, worms, molluscs, sea urchins, starfish, lamp shells and trilobites. The various forms of life of that time, like the organisms of today, can be classified as belonging to three 'domains' — namely, the Archaea, the Bacteria and the Eukarya. The cells of Eukarya contain nuclei, and this domain includes Protista (e.g. amoebae), Fungi, Plants and Animals.

The next 500 million years

Five hundred million years ago, there were animals swimming in the oceans that had an internal supporting structure or backbone. The earliest of these were the so-called jawless fishes, a group represented today by the lampreys. By 400 million years ago, the so-called 'true fishes' were just emerging, although the oceans were dominated by arthropods, especially trilobites.

There was much less diversity among plants. At that time all plants were thallophytes, which exhibited no real differentiation into stems, leaves and roots. This group included various kinds of multicellular algae, like stoneworts and brown seaweeds.

The main plants of the oceans have changed little since that time. In contrast, spectacular evolutionary changes took place among animals in the aquatic environment. By 200 million years ago the trilobites, which had dominated the scene for so long, had entirely disappeared and were replaced by a new group of molluscs known as ammonites. At one time there were over 20 different families of ammonites, and some of them had a diameter of at least a metre. But the ammonites were also extinct by 60 million years ago.

Meanwhile, there was remarkable diversification taking place among the bony fishes, leading eventually to the immense variety of fish species that are found in ponds, streams, rivers, lakes and oceans today.

The earliest plants to grow on land appeared on the edge of the shallow water of estuaries a little over 400 million years ago. Unlike the thallophytes in the oceans, the earliest land plants had a distinct stem that provided them with support in the new environment, and some of them had rudimentary leaves. Fossilised remains have been found of two distinct groups related to the modern psilotums and club mosses. Eventually larger plants evolved. By 350 million years ago, there were great forests of seed ferns and horsetails. Because their reproduction depended on the sperm being able to swim in a film of moisture to reach the ovum, they could exist only in moist areas. This is still the case today for the mosses, liverworts, psilotums, horsetails, ferns and club mosses. Seed ferns are now extinct.

The colonisation of drier land by plants depended on the evolution of a means of reproduction that did not require the sperm to swim through a film of water. This came about in the development of a pollen tube through which the sperm passes to reach the ovum. The first plants with a pollen tube were the gymnosperms, which appeared around 300 million years ago. There were four main kinds of gymnosperms — the cordaites, which are now extinct, and the cycads, ginkgos (maidenhair trees) and conifers.

It was also around 400 million years ago that the first animals ventured onto land. Except in the case of worms, this development involved some important structural changes that enabled them to resist drying out, to breathe atmospheric oxygen, and to move around from one place to another. The first of these requirements was met by the formation of a resistant outer skin, and the second by the development of cavities in the body into which air could pass and from which oxygen could be transported to the various tissues. Locomotion on land in the crustaceans, centipedes, spiders and, later, insects was made possible by modification of the limbs that already existed in earlier aquatic forms. The five-toed limbs of the vertebrates evolved directly from the fins of their fish ancestors.

The heyday of the amphibians was around 300 million years ago, when many diverse forms existed. By 200 million years ago, however, their numbers had declined dramatically, and their place had been taken by reptiles, including the earliest dinosaurs. Birds and mammals evolved directly from reptiles.

Reptiles, including the dinosaurs, showed extraordinary diversification, with different groups becoming adapted to many different kinds of habitat. Several aquatic groups evolved, some of which looked very much like fish, although they did not have gills and they breathed air through a respiratory tract. There were also various forms of flying reptiles, with wings spanning up to seven metres and made of leathery membranes, supported and extended by elongated fingers.

Between 60 and 70 million years ago, a great crisis occurred in reptilian history and many forms became extinct, including all the dinosaurs and flying reptiles and most of the large marine reptiles. Many other forms of life disappeared during this period of reptile extinction, including various microscopic foraminifera in the oceans and many aquatic animals, including the ammonites. Whatever the cause of this wave of extinction, placental mammals, birds, lizards, snakes, turtles, crocodiles, fishes and plants were relatively unaffected.

The earliest mammals came into existence about 200 million years ago, at about the same time that the dinosaurs were emerging as a distinct group; and there were animals very like modern echidnas wandering around 150 million years ago. But mammals remained a rather insignificant group during this period of reptile dominance.

The evolutionary transition from reptiles to mammals involved three especially important changes. First, except in the case of the egg-laying platypus and echidna, a mechanism evolved by which the embryo developed within the mother's body, attached to maternal tissue by a placenta through which oxygen, carbon dioxide, nutrients and waste products passed to and from the embryo. A somewhat similar arrangement is found in a few reptiles, such as the Australian blue-tongue skink. Second, in all mammals the newborn young are cared for by the mother and nourished by milk from her mammary glands. Third, a mechanism developed in mammals and birds that maintained a more or less constant body temperature, relatively independently of muscular activity and environment. It has been suggested that similar mechanisms may have existed in dinosaurs.

The first true flowering plants, the angiosperms, emerged about 160 million years ago and, since that time, they have undergone spectacular diversification. They are now the dominant division of plants. They are made up of two main groups — the monocotyledons and dicotyledons. The seedlings of the monocotyledons, which include grasses, lilies, irises and crocuses, have a single leaf and the stems do not thicken. The seedlings of dicotyledons have two leaves and the stems become thicker as the plant matures.

After about 60 million years ago, an amazing evolutionary diversification took place among birds and mammals. The primates, for example, which had emerged during the last part of the dinosaur era, evolved into four main groups. The most ancient group is the prosimians, which includes lemurs, aye-ayes, lorises and tarsiers. The second group, the ceboids, consists of the monkeys of South America. These animals have tails by which they can hang from branches of trees, and the group includes marmosets, howler monkeys and spider monkeys. The third group is the ceropithecoids, the monkeys of Africa and southern Asia, and it includes baboons, mandrills, langurs, and macaques. These animals also have tails, but they cannot hang by them. The fourth group, the hominoids, which includes gibbons, orangutans, chimpanzees, gorillas and humans, do not possess tails.

The evolutionary sequence of life on Earth is depicted diagrammatically in Figure 2.2.

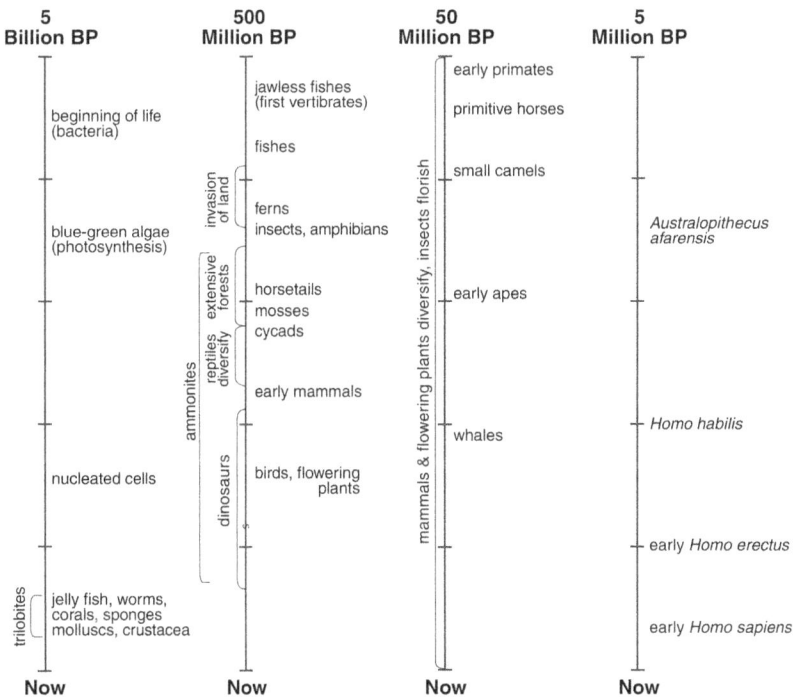

Figure 2.2 Some major developments in the history of life
Source: Stephen Boyden

The mechanism of evolution

According to the Darwinian explanation, evolutionary change comes about through natural selection. This process depends on the fact that, at any given time, the individuals in a population of living organisms are not genetically identical. This genetic variability is due partly to changes, or mutations, that occur spontaneously from time to time in the genetic material of the sex cells (gametes), and partly to the fact that genes occur in different combinations in different individuals.

Because the members of a population are not genetically the same, some of them are likely to be better suited than others to the prevailing conditions. These better suited individuals tend to be more successful in surviving and reproducing, and are therefore likely to contribute a greater number of individuals to the next generation. Their progeny will carry the genes that rendered their parents at a biological advantage. Consequently, generation by generation, a population can become increasingly well suited to the environment in which it lives.

Similarly, when a significant and lasting change occurs in the environment of a population, some individuals, because of the genetic variability in the population, may be better suited than others to the new conditions. These individuals are more likely to survive and successfully reproduce, passing on their genes to subsequent generations.

Not all populations adapt successfully in this way to environmental change. Indeed, the great majority of species that existed in evolutionary history eventually failed to adapt to new environmental conditions and became extinct.

The rate at which evolutionary adaptation occurs in a population following environmental change depends on a number of factors. Especially important among these is the frequency in the initial population of 'favourable' genes associated with resistance to the threats inherent in the new situation, and the extent to which such genes confer an advantage on the individuals that carry them (i.e. their selective advantage).

The mutation rate for individual genes is estimated to be around one mutation per 100,000 spermatozoa or ova, and most mutations are harmful rather than beneficial. The chances of a suitable or helpful mutation arising in an appropriate gene in a small population that is suddenly exposed to a new detrimental environmental condition are, therefore, negligible. In the long term, however, all major evolutionary change depends on the introduction of new genetic characteristics through random mutation.

Extinctions

The fossil record shows that the evolution of life has not been an entirely gradual process. There have been five periods of mass extinction resulting from major changes in global conditions. These occurred around 450 million years ago, 360 million years ago, 250 million years ago, 205 million years ago and 65 million years ago.

In the most severe of these mass extinctions — the one that took place about 250 million years ago — 57 per cent of all families became extinct, as did 83 per cent of all genera and 96 per cent of all species.

After a wave of extinction, many ecological niches are left vacant, and this encourages relatively rapid evolutionary change and diversification among surviving populations. This is what happened after the disappearance of the dinosaurs 65 million years ago. Before many millions of years had passed, the ecological niches that these creatures had vacated were occupied by new kinds of animals.

Energy and ecology

The rich diversity of plants and animals that exist in the modern world, including humans and their civilisation, could not exist without photosynthesis.

Plants use about half the energy they capture from sunlight in their own vital processes, eventually releasing it into the environment in the form of heat. The remaining energy takes one of several pathways. Dead plant tissue containing stored energy is broken down by bacteria or fungi, which make use of the energy in their own vital processes and eventually release it into the environment as heat. Some plant tissue is consumed by plant-eating animals, providing them with the energy they need for their life processes. Some of this energy is given off in the form of heat, while some of it is retained in the animals' own tissues, eventually to be consumed either by carnivores or by microorganisms, and ultimately returned to the environment in the form of heat. This sequence of events is referred to as a food chain, with plants playing the role of producers, animals the role of consumers, and microorganisms and fungi the role of decomposers (Figure 2.3). Sometimes the chemical energy stored in plants is converted directly to heat through the action of fire.

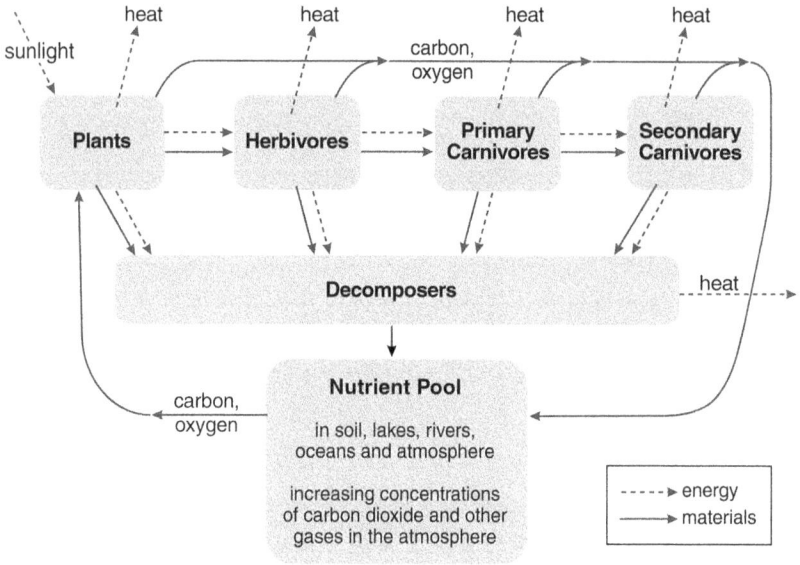

Figure 2.3 The nature of food chains
Source: Stephen Boyden

A very small fraction of dead plant tissue avoids these various fates and only partially decomposes. Under certain conditions, like those that are likely to exist in swamps or bogs, decomposition of dead plant material may be incomplete due to lack of oxygen in the stagnant water and acidity resulting from the decay process. The soft, fibrous energy-containing material formed in this way is called peat. Downward pressure resulting from the accumulation of sediments above may eventually transform peat into coal.

Petroleum and fossil gas, which are also of organic origin, are produced by the breakdown of vast quantities of microscopic plants and animals in the oceans. Unlike coal, the liquid and gaseous hydrocarbons often migrate from their place of origin to become concentrated in distant reservoirs. The formation of the deposits of these fossil hydrocarbons spanned several hundred million years. They are now being used by humans as sources of energy at a rate that is several million times faster than the rate at which they were formed.

Nutrient cycles

While the energy on which life processes depend comes from outside the biosphere in the form of light and is eventually returned to outer space as heat, the material components of living organisms come from the planet itself. An essential characteristic of life on Earth is the cycling within the system of chemical elements that are taken up from the environment, built into the tissues of living organisms, and then eventually released again into the environment — to become available for incorporation into new life.

Plants take up the various nutrients that they need for their growth from their immediate environment. Carbon is taken from the atmosphere and oxygen from the atmosphere, soil and water. All other essential nutrients are taken from the soil and water.

These nutrient cycles are essential for the sustainability of life in all natural terrestrial ecosystems.

The different nutrient cycles vary in complexity. The carbon and oxygen cycles are intimately connected and are relatively simple (Figure 2.4). The nitrogen cycle is more complicated. Plants take up the nitrogen that they need for growth from the soil and water in the form of sodium nitrate. This is made available largely through the activities of certain bacteria in the soil, some of which manufacture it from breakdown products of decomposers, and some by fixing free nitrogen from the atmosphere so that it becomes assimilated into organic compounds.

Indeed, the whole of multicellular life on Earth is ultimately dependent on the activities of microbes because of the essential roles that they play in the breakdown of the tissue of dead plants and animals and in the cycling of nutrients.

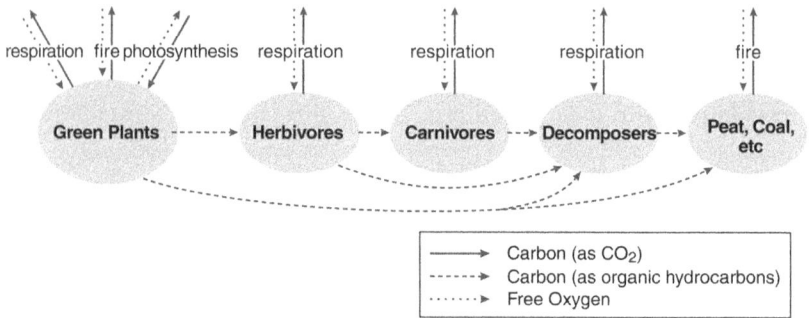

Figure 2.4 Carbon and oxygen cycles
Source: Stephen Boyden

Ecosystems

The term ecosystem is applicable to any system of interconnected living organisms and the physical environment with which they interact, from a small pond or vegetable garden to a forest, a continent or the biosphere as a whole.

At a basic level we can recognise two broad classes of ecosystem, aquatic and terrestrial, although many ecosystems incorporate both aquatic and terrestrial elements. The biological characteristics of ecosystems are largely determined by their latitude and altitude and by temperature and rainfall.

Terrestrial ecosystems are conventionally classified as follows: desert; tropical grassland and savannah; tropical scrub forest; tropical rainforest; temperate grassland; temperate forest (deciduous, or eucalyptus and acacia); northern coniferous forest; and tundra (a treeless zone lying between the ice cap and the timber line in the northern hemisphere and also at the southern tip of South America). Cities can also be seen as ecosystems, although they are totally dependent on inputs from ecosystems elsewhere for their survival.

Individual species in an ecosystem are said to occupy a particular ecological niche, a term that refers not only to the physical space in which a species lives, but also to its functional role in the ecosystem.

Soil

Soil has been defined as the unconsolidated portion of the Earth's crust that supports plant life, and it is made up of debris resulting from the weathering of rocks as well as organic matter. The fragments of rock may range in size from largish lumps, through sand, down to fine clay.

The chemical composition of soil influences plant growth and is largely determined by the kind of rock from which it is formed. Soil quality is also affected by the vegetation that it supports. Thus, the soil and vegetation develop together, each influencing the other, and each being influenced by, and to some extent influencing, the climate.

The organic content of soil consists of decomposing plant and animal matter, and the living microorganisms involved in this decomposition. It also contains numerous other microorganisms, such as those participating in the fixation of nitrogen from the atmosphere. Some components of decaying organic matter, like waxes, lignins and fats are relatively resistant to decomposition, and together they form a colloidal substance known as *humus*. Humus has an important influence on the capacity of the soil to support plant life.

Soil also contains many animals, including nematodes, millipedes, mites, insects, earthworms and burrowing amphibians, reptiles and mammals. There are believed to be around 1 million species of nematodes, most of them living in soil.

The living organisms in soil play a crucial role in the nutrient cycles on which all life depends. Although the organic component of soil may represent less than 0.1 per cent of the total soil mass, it can still amount to several tonnes per hectare. In the case of grassland, for example, the weight of organic matter in the soil in a given area is many times the weight of cattle, sheep, kangaroos or other herbivores that could be supported on that land.

Uniformity and diversity in nature

We have only to look around us anywhere in the natural environment to be struck by the amazing diversity among living organisms — diversity in habitat, size, shape and colour; and, among animals, diversity in means of locomotion and patterns of behaviour.

Animal species also differ widely in their food sources and in their resistance to heat, cold, dryness and wetness; some are at home on the land, some in the water, some in the soil, and some in the air. Each is adapted, in its inheritable characteristics, to its particular ecological niche.

It is impossible to state precisely how many different kinds of life now exist on Earth, but it has been roughly estimated that there are some 7 to 10 million species of eukaryotic organisms (i.e. all organisms excluding bacteria, archaea and viruses).

Uniformity

Underlying all this diversity, however, there are some remarkable and essential uniformities. One of these fundamental universals is the fact that all forms of life depend on a continual supply of energy. Except in the case of a small proportion of microbial organisms, this energy was initially captured from sunlight by photosynthesis in green plants, converted into chemical energy and stored in organic molecules.

There is also a basic similarity in the complex chemical processes by which this energy is used in living cells, be they animal, plant or microbial. For example, a common denominator at the molecular level is adenosine triphosphate (ATP). In every kind of living organism this substance plays an essential part in the chemical reactions involved in the storage of energy and its eventual release, for instance, in the synthesis of complex molecules or the contraction of muscles.

The organic molecules of which organisms are made up also share the same basic characteristics. These molecules fall into four classes: carbohydrates, proteins, lipids (fats) and nucleic acids. Within these four main classes, however, there is immense diversity. In the case of proteins, for example, every species of animal and plant has many different proteins with different functions, like enzymes and hormones, or playing specific structural roles; and the proteins of each species are distinguishable from those of all other species. Indeed, subtle differences exist in the structure of proteins between individual members of the same species. This is why skin, or other organs, can only rarely be successfully grafted from one individual to another (except in the case of identical twins) unless special steps are taken to depress the immune system of the recipient. The rejection of a tissue graft from another individual is due to the fact that the immune system recognises the cells of the donor as 'foreign', and consequently sets up an inflammatory response, which ultimately destroys them.

Another universal is the fact that all life forms, with the exception of sub-microscopic viruses, have a cellular structure — ranging from single-celled organisms, like bacteria and amoebae, to the large multicellular plants and animals, which may be made up of hundreds of billions of separate cells with many different functions. But every one of these multicellular animals, and most of the multicellular plants, begin life as a single cell, formed by the union of two cells, the ovum and the sperm.

Also universal among cellular organisms is the means by which genetic information is passed from parents to their progeny, providing the instructions that result in the new organisms developing and functioning as members of the species to which their parents belong, and that determine all their other inherited characteristics.

The essential agent in this process is the genetic material of the cell, deoxyribonucleic acid (DNA). In animal and plant cells, chains of DNA are located in the cell nucleus and, in this situation (and, in some laboratory situations, outside the living cell), DNA is itself capable of self-replication. It contains, in coded form, most of the information that is necessary for the formation of the new individual.

The inheritable characteristics of all organisms are determined by the arrangement of four nucleotides (cytosine, thymine, adenine and guanine) in the genes, which are discrete areas or regions on the DNA chains.

Almost universal among plants and animals is the involvement of the sexual process at some stage in the reproductive cycle. This consists of the fusion of two separate cells (gametes) that, in the case of multicellular organisms, usually come from two different individuals, although in some species, they come from different parts of the same individual. In some very simple organisms the two gametes may be identical, but in all higher species of plants and animals they are clearly different. One, the male gamete, or sperm, is motile. The other, the female gamete, or ovum, is larger and sessile. The fusion of the two cells results in the new fertilised egg, or zygote, which contains twice the amount of DNA contained in each of the gametes. A mechanism known as meiosis results, however, in the amount of genetic material being halved, so that the total amount of DNA does not double at each generation.

The fertilised egg thus contains genetic material from two different individuals. Since it is unlikely that the material from each parent will be identical, it follows that the offspring will be genetically different, even if only slightly, from either parent.

The sexual process means that the genetic material in a population is being constantly reshuffled. From the evolutionary point of view, the importance of sexual reproduction lies in the fact that, unlike in asexual reproduction, the precise genetic make-up of the new individual is different from that of either parent. This has the effect of maximising the number of genetic combinations in the population, and so increasing the potential of the population to adapt to environmental change through natural selection.

While the mechanism of sexual reproduction explains the continual rearranging of genetic material in populations, it does not tell us how entirely new genetic characteristics come into existence. This happens through the process of mutation, which consists of a chemical change in a gene that is perpetuated when the gene replicates in cell division. The change then affects the particular characteristic of the organism for which the gene is responsible. Mutations are normally rare events, but their frequency can be increased by certain physical and chemical agents, such as ultraviolet light, radioactive radiation and mustard gas.

The great majority of mutations are deleterious, so that cells that carry them do not survive. Occasionally, however, a mutation arises that, by chance, increases the likelihood of the organism surviving and successfully reproducing in the habitat in which it lives.

Sometimes reproduction occurs without the involvement of a male organism. In this process, which is known as parthenogenesis, a new individual develops from an unfertilised egg. It is not uncommon among plants but rare in animals, although it has been recorded in some invertebrates, including nematodes, water fleas, aphids, some bees and parasitic wasps and in a few vertebrates, including some lizards, geckos and fish.

Diversity

Despite these fundamental uniformities, evolution has given rise to a fantastic variety of structural forms, physiological mechanisms and ways of life. Let us look at just a few examples to illustrate the extent of this diversity, focusing on two key aspects of life — food intake and reproduction.

Food intake in plants

The range of ecological niches exploited by plants is vast, and very different forms of vegetation can be found in deciduous and coniferous forests of the northern hemisphere, dense evergreen forests of the tropical zones, eucalyptus forests of Australia and in mountainous terrains, savannah, marshlands, deserts, heathlands and sand dunes across the globe.

Each plant form is adapted, through evolution, to certain conditions of temperature, humidity, soil quality, soil wetness, light and wind.

While some water is essential for the survival and growth of all plants, enormous variation exists in the amount of water that different plants need. Some forms, like most reeds and bulrushes, cannot survive in soil that does not have a high water content, while others are adapted to extraordinarily dry conditions. Plants found in dry habitats often have small, leathery leaves. An extreme example is the desert cacti in which the leaves are hard, spiny structures that do not support photosynthesis. In these plants the photosynthetic process takes place in the fleshy stems, which are also organs for storing water. Their water content may account for up to 98 per cent of their weight.

There are many other kinds of adaptation to dry conditions in plants. One of these takes the form of short life cycles. Parts of the Australian desert may receive a reasonable rainfall only once in every few years. When this occurs, the previously parched and apparently lifeless ground suddenly becomes a mass of small flowering plants and, in a very short time, seeds are produced. If there is no further rain, the soil soon returns to a state of desiccation that contains myriads of drought-resistant seeds lying dormant until the next time it rains.

In most leafy plants the size of the pores, or stomata, on the leaves varies in response to changes in the moisture content of the soil and the humidity of the atmosphere, so controlling the rate of water loss by evaporation. In some plants that live in dry regions, the stomata are permanently sunken into the surface of the leaf, thus minimising evaporation, while in others the leaves are covered with hairs that have the same effect. In many plants the leaves fold up when conditions become dry and, in some forms, the leaves fold regularly after dark and sometimes in the late afternoon. In most plants only about 1 or 2 per cent of the water taken up by the roots is used in photosynthesis and the rest is released by the stomata into the atmosphere, through the process of transpiration.

A particularly interesting adaptation to nutrient deficiency in soils is seen in the carnivorous plants, of which there are at least 350 different species. These plants are usually found in swamps, bogs and peat marshes where acids have leached the soil of nutrients. Their prey may consist of insects and other invertebrates and sometimes even small birds and amphibians. Sundews, for example, are very small plants, usually not more than five centimetres across, and they have tentacles on the upper side of the leaf that secrete a clear sticky fluid to attract insects. As soon as an insect is caught by one tentacle, the others bend inwards towards it, so that the animal is thoroughly trapped. The tentacles then secrete enzymes to digest the insect tissues, and the soluble nutrients are absorbed by the leaf surface. Among other carnivorous plants is the well-known Venus flytrap, which occurs naturally on the coastal plain of North and South Carolina, in North America. Unlike carnivorous animals, carnivorous plants do not use their prey as a source of energy, but rather as a supplementary source of nutrients, especially nitrogen and phosphorus.

A more common way of acquiring nitrogen operates in legumes, such as clovers, vetches, lucernes, peas, beans, and acacias, and involves a symbiotic relationship between the plant and certain nitrogen-fixing bacteria known as rhizobia. When the plants are seedlings their root hairs are invaded by the rhizobia, and eventually these give rise to small nodules in which the bacteria live and multiply. These microorganisms fix free nitrogen and release it in the form of ammonia, which combines with carbon compounds in the plant cells to produce amino-acids. In agricultural systems the beneficial effects of growing legumes has been appreciated for over 200 years. Some of the fixed nitrogen is released into the soil around the legumes and so becomes available to other plants. If leguminous plants are ploughed back into the soil, much of the incorporated nitrogen becomes available for other crops. A crop of lucerne ploughed back into a field may add as much as 350 kilograms of nitrogen to the soil per hectare.

Food intake and digestion in animals

Turning to the procurement and assimilation of food in animals, the basic arrangement of the digestive tract — that is, a single mouth, a stomach and intestines containing digestive juices, and a single anus — is common to all multicellular animals, from mosquitoes to elephants, with the exception only of some simple forms, like sponges and flatworms. The extent of variation on this common theme, however, is enormous.

First, the great range of different kinds of food sources has resulted in wide variation in the structure of the mouth parts in different animals. Some examples are the grinding molars of herbivores (e.g. ox, horse); the sharp cutting teeth of carnivores (e.g. dog, tiger); the beaks of sparrows and pelicans; the sucking mouthparts of leeches; the powerful biting and chewing jaws of the praying mantis; the proboscis of mosquitos; the fly-catching tongue of chameleons; and the simple oral cavity of earthworms.

There is also great diversity in the various organs concerned in the digestive process and in the biochemical properties of the digestive juices. Because of the specificity of these adaptations, if animals are forced to consume a diet that is significantly different from that to which they are adapted through evolution, they are likely to show signs of ill health. Tigers will not last long on a diet of honey, and bee larvae could not survive on a diet of meat.

The following few examples illustrate the range of adaptations in the internal digestive organs. Termites eat mainly wood. Like other animals, however, they do not produce any enzymes in their digestive tracts that are capable of breaking down the lignin of which wood is made. They are entirely dependent for their nutrition and survival on certain microorganisms that live in their stomachs and which produce an enzyme to split lignin into soluble carbohydrate molecules that can be utilised by the termite.

There is wide variation in the structure and physiology of the gastro-intestinal tract among birds. In most species, the lower end of the oesophagus swells into a large storage chamber, the crop, where the food remains, sometimes for as long as two days, until the stomach can accommodate it. Crops are prominent in many grain-eating birds, allowing them to swallow a large volume of food in a hurry, so shortening their time of exposure to predators. In pigeons, the crop takes the form of a large double sac that not only stores grain, but which also secretes 'pigeon's milk' for feeding the young birds.

In grain-eating birds, the stomach itself consists of two parts, the anterior glandular stomach, which secretes digestive juices, and the posterior muscular stomach, or gizzard. The gizzard is especially well-developed in grain-eating birds and is lined with horny plates or ridges that serve as millstones for grinding the food. This process is often furthered by the abrasive action of small pieces of grit that the birds have swallowed. The gizzard of the domestic goose may contain 30 grams of grit.

In carnivorous birds, the gizzard usually has much thinner walls and has a completely different function. In owls, gulls, swifts, grouse and some hawks it operates as a trap that stops sharp bits of bone and other non-digestible fragments from passing on through the alimentary canal. This material is rolled up into elongated 'pellets' which are regurgitated through the mouth.

A further example of an alimentary adaptation to a specific kind of diet is provided by the four 'stomachs' of ruminants such as cattle and giraffes. These animals tear the leaves off the plant they are eating with their incisors and swallow them almost immediately, without making any attempt to chew them up. The food bypasses the 'first stomach', or rumen, and goes directly to the smaller 'second stomach' or reticulum, where it is compacted into balls. At a later time, when the animal has stopped feeding, these balls, known as the cud, are regurgitated to the mouth. The cud is then properly chewed by the grinding action of the animal's molars, before being swallowed a second time, this time to be retained in the rumen. This large organ represents about 80 per cent of the total volume of the four stomachs. It is colonised by bacteria and protozoa, which not only break down cellulose, as in termites, but also synthesise proteins, using urea and ammonia as nitrogen sources. Some of these microorganisms pass down the alimentary canal and are themselves digested, so contributing to the animal's intake of amino-acids. Some of the products of the fermentation are absorbed directly by the lining of the rumen. The rest of the food passes into the omasum, or third stomach, which basically functions as a strainer, and then on to the abomasum. This is the true stomach, where digestive enzymes are secreted. Anatomically, the rumen, reticulum and omasum are actually expansions of the oesophagus.

There is also a great deal of variation among animals in the ways that they find and procure their food. Many species locate their food simply by going around looking for it, in much the same way as we would ourselves, using especially the senses of sight, smell and hearing. Clearly, there is a broad distinction between the techniques of herbivores and carnivores, in that carnivores, except in the case of scavengers, have not only to locate their food source, but also to catch it. Some groups of animals have specialised modes of food location and procurement. Bats, for instance, have evolved the mechanism of echolocation for detecting their prey in the night sky. The technique involves the emission of high frequency sounds and the detection, by means of highly specialised listening devices, of echoes of these sounds coming from objects in the environment. When the returning signal indicates that the object detected is of an appropriate size and is moving in the air, the bat flies rapidly and unerringly towards it, and catches it. The bat is able to discern from the signal whether the object is flying towards or away from it. A similar mechanism has evolved independently in dolphins, which also emit ultrasonic pulses, and the pattern of returning echoes provides them with a picture of the world around them.

In some carnivorous animals that feed in water, receptors have evolved that detect small electric impulses generated by the muscular movements of their prey. The platypus, which is effectively blind under water, detects small crustaceans and worms in this way. Frog tadpoles and some fish make use of similar mechanisms.

Any discussion about food acquisition in animals would be incomplete without reference to the farming practices of certain ants that live in tropical and sub-tropical regions on the American continent. Some of these species collect pieces of leaves or flowers from living plants and carry them back to the nest, where they cut them up into smaller pieces and mix them with saliva and faeces. The ants spread out the resulting compost in an underground garden, and then place pieces of mycelium from a certain kind of fungus on top of it. The fungus grows profusely, deriving nourishment and energy from the cellulose in the leaves or flowers. As the mycelium grows, the ants continually make cuts in it and, at the site of each cut, the fungus develops a nodular proliferation. These nodular proliferations are eventually harvested by the ants as a major food source. Some other ants in the same region make use of this principle, but use insect faeces or dead insects as a substrate for the fungal mycelium.

Reproduction

The ability to reproduce and so perpetuate the species is, of course, an essential feature of all forms of life.

Despite the underlying uniformities at the molecular level, the details of the processes of reproduction at the level of whole organisms vary enormously. First, let us note the all-important distinction between sexual and asexual reproduction. In asexual reproduction there is only one parent, which splits, buds or fragments to give rise to two or more new individuals, each of which have hereditary characteristics identical with those of the parent. Asexual reproduction is common among simpler forms of life, including bacteria, algae, fungi, mosses, protozoa, coelenterates and flatworms. In the case of the last group, if the animal becomes fragmented into several pieces, each may develop into a new whole animal. If a starfish is cut in two, each part will regenerate tissue to form a complete new starfish.

Many plants are capable of reproducing asexually. Some species, such as English elms and Lombardy poplars, may propagate by putting out 'suckers', so that new trees grow up from the distal roots of the parent trees. Reproduction by rhizomes or 'creeping rootstocks' (actually stems growing laterally underground) is common — as in bamboos, hops, asparagus, irises and many other plants. Some species, like potatoes, reproduce asexually by means of tubers.

Propagation of plants by means of cuttings is another example of asexual reproduction.

Indeed, asexual reproduction also occurs in higher animals, including humans, when a newly fertilised egg divides in the uterus to give rise to two or more genetically identical eggs, each of which develops as an independent organism. Today, as an outcome of scientific advances, it is now possible to bring about asexual reproduction artificially in mammals by means of cloning techniques.

Turning to sexual reproduction, we have already noted some basic differences between the simpler, more ancient plants, like mosses and ferns, and the more recent conifers and flowering plants. Let us now look at a few of the adaptations that have evolved in this last group.

The most striking feature of the reproductive processes of the flowering plants is the fact that, while wind sometimes plays a part in transporting pollen from flower to flower, the great majority of species rely entirely on insects or, in some cases, on small birds or mammals, to bring about pollination. For this to work, the insects have first to be attracted to the flowers so that they pick up pollen and later drop it off when they visit other flowers of the same species, where it can bring about fertilisation. The basic attractant for insects in the great majority of plants is food, in the form of nectar, which is produced at the base of the flower solely for this purpose. Another feature of the adaptation of the flowering plant is the development of petals, which are often displayed conspicuously and in bright colours, signalling the presence of nectar.

While this basic pattern is very common, there are many interesting and sometimes bizarre variations on the general theme. As Darwin noted in his remarkable book on orchids, these plants are especially interesting. In one species the shape and colour of the flower bears a strong resemblance to the female of a particular species of wasp, complete with eyes, antennae and wings. It even gives off a similar odour to that emitted by a female wasp that is ready to mate. Male wasps, deceived by this arrangement, attempt to copulate with the flower. In doing so, they pick up pollen, which they inadvertently deposit on the next flower that they mistake for a female wasp.

The dung lily is another interesting example. This plant gives off an odour similar to that of herbivore dung. When a dung beetle happens to fly overhead, it responds to the dung-like stimulus by dropping head first into the funnel-shaped flower. Because the inside of the flower is lined by small hairs pointing downwards, the beetle is unable to climb out and, if it happens to be carrying pollen from a previous encounter with a dung lily, some of this will come off and fertilise the ova. By morning, the flower tips over and the one-way hairs no longer prevent the beetle from escaping. On the way out, the beetle picks up additional pollen that was not available when it entered.

In multicellular animals, two main mechanisms exist for achieving union of egg with sperm. The first operates only in the case of animals that live, or at least mate, in water, and it involves the male liberating sperm into the water in the region where the female has recently laid her unfertilised eggs. Usually this is preceded by certain courtship behaviours to ensure that the male is at the right place at the right time. This method operates in most marine animals, from molluscs to true fishes, as well as in amphibians, which return to the water to mate. In frogs the male arranges himself on the back of the egg-laden female, keeping firmly in place by means of special clasping pads on the front of his forelimbs. He remains in this position until the female begins to lay her eggs, at which time he ejects spermatozoa into the water, some of which find and unite with ova.

The pattern in newts and salamanders is somewhat different. In the common newt of north-western Europe, *Triturus vulgaris*, mating takes place in the water and the male courts the female with a dance display involving a rapid waving movement of the end of his tail, which is turned back on itself, and so points forward. When the female is appropriately aroused, apparently partly as a result of a hormone discharged into the water from the male cloaca, the male newt deposits a mucilaginous bundle of spermatozoa, which the female picks up with her hind limbs and inserts into her cloaca, so that fertilisation takes place internally.

The main mechanism in land animals for bringing sperm and eggs in contact involves the insertion of the penis of the male into the genital tract of the female, followed by the ejection from the penis of spermatozoa which then swim their way to the ova. This mechanism exists in most insects, in some birds, and in all reptiles and mammals.

Different procedures operate in worms and some arthropods. The reproductive pattern of the earthworm is particularly complicated. These animals are hermaphrodites and during mating the two worms, heading in opposite directions, lie with their ventral surfaces in opposition and are held together by a sticky secretion. Each worm donates sperm to the other, and these are temporarily stored in a seminal receptacle. After the worms have separated, a glandular ring of thickened skin, the clitellum, secretes a membranous cocoon. As the worm frees itself from this cocoon, it discharges into it the ova produced in its own body as well as the sperms contributed by the other worm. As the cocoon slips off the worm its two openings constrict, and the fertilised eggs then develop inside it to produce new worms.

Another interesting mechanism has been observed in certain species of peripatus — caterpillar-like animals that live in moist forests in Africa, Asia, Australia and South America. They have many pairs of legs (13 to 43 pairs depending on the species), and they share characteristics of both the annelid worms and arthropods. In some species of peripatus, males have a special protuberance on their head that is used to carry around a drop of semen while it searches for a female. When a female is found, the male deposits the semen somewhere on the surface of her body, which causes a reaction to take place inside the female and, as a result, some specialised cells in her body transport the sperm to the ova in the uterus.

Reproduction in spiders is rather similar to the first part of this peripatus procedure. The male spider produces a ball of sperm-containing material that he picks up with one of his pedipalps, which are limb-like structures situated just in front of his four sets of legs. He then sets out in search of a female, which, in the case of most species, he must approach with considerable caution, identifying himself by certain species-specific signals in order to avoid being attacked and eaten. On reaching the female, the male inserts the spermatozoa into the female genital tract. In most cases, he then quickly makes his getaway although, in some spider families, the female consumes the male as soon as mating is completed.

A great variety of procedures exist among different animals for ensuring that males and females find each other for mating purposes. In many instances, the female gives off a specific odour that attracts males. In some moths, the males are exquisitely sensitive to these odours, responding when there are only about a hundred molecules of the specific substance per millilitre of air. It has been estimated that, in some kinds of moth, the male can detect a female over 4,000 metres away if a gentle breeze is blowing in the right direction.

In other species, the male attracts the female to a particular place or territory by emitting a distinctive call. This pattern is common among birds and frogs. In some bird species, the peacock and the Australian lyrebird being notable examples, males attract females by extending and displaying their tail feathers. In bowerbirds, the males achieve the same objective by constructing a 'bower' and decorating it with all sorts of colourful objects.

In the great majority of mammals, from rats, mice and shrews, to dogs, zebras, elephants and monkeys, females undergo a hormonally controlled cycle during which there are certain periods coinciding with ovulation when they are sexually attractive, or receptive, to males. The important result of this mechanism is that mating takes place only at times when fertilisable ova exist in the female genital tract. An outstanding exception to this generalisation is *Homo sapiens*, in which females can be sexually attractive to males at all times, and in which female receptivity is not restricted to a short period in the hormonal cycle.

In all mammals, milk produced in the mammary glands of females is, with one exception, the only source of food for their newborn offspring. Only one species is known — the guinea pig — in which newborn animals can survive without milk, eating solid food immediately after birth. Newborn guinea pigs, however, do drink milk from their mothers if it is available. At the other extreme is the young grey kangaroo, which weighs less than one gram when it is born and which, despite making its own way from the urogenital opening of the mother to the pouch, is otherwise completely helpless. Once in the pouch it immediately becomes attached to one of the nipples, and it does not leave the pouch, even for short periods, for nine months.

Comment

The examples given above only touch the surface of the vast range of different life forms that exist on Earth. The shelves of science libraries hold countless volumes providing detailed information on the structural, physiological and behavioural diversity encountered among living organisms. And, apart from all that has already been described, there is much more yet to be discovered.

Health and disease

For any living organism, health can be defined as follows:

> Health is that physiological and behavioural state most likely to ensure survival and successful reproduction.

In the case of animals, health is thus consistent with optimal performance in terms of procuring food and water, avoiding predators, mating, giving birth and, in many species, successfully raising young. Health is a relative concept, however, in so far as the state of an organism can be anywhere on a continuum from optimum health at one extreme to near death at the other.

The health needs of animals, including *Homo sapiens*, are determined by their evolutionary background. This is because, through the processes of evolution, species have become well adapted in their innate biological characteristics to the conditions prevailing in the environment in which they are evolving. It follows that these conditions are capable of satisfying their health needs.

If an organism is exposed to conditions of life that differ significantly from those that prevail in its natural environment — that is, the environment in which it evolved — it is likely to be less well adapted to the new and different environment, and it is therefore likely to show signs of maladjustment. It will be less healthy than in its natural environment. This fundamental evolutionary health principle applies both to plants and animals.

Thus, in the case of plants, all species are biologically adapted to the conditions prevailing in the environment in which they evolved. That is, they are adapted to certain kinds of soil (e.g. depth, chemical constitution and water content), a certain intensity and quality of solar radiation, certain atmospheric conditions and a certain temperature range. If they are exposed to conditions that differ significantly from those to which they are adapted, they will not grow well or will die.

The evolutionary health principle is taken for granted by those responsible for the health of animals in zoos. Zoo keepers try to provide creatures in captivity with the same kind of food that they normally eat in the wild and, if possible, to ensure that they are exposed to temperatures similar to those of their natural habitat.

If, like hippopotamuses, they naturally spend most of their time in water, they will be provided with water to wallow in. If the animals in the wild live in trees, then they will be provided with branches to climb. In other words, zoo keepers appreciate that the best guide to the health needs of any species is information about their conditions of life in the environment to which they are biologically adapted through evolution.

Clearly, the natural environment does not satisfy the health needs of all creatures all of the time; every animal eventually dies. But, in animal populations in their natural habitats, most of the individuals are in a state of good health most of the time. This applies to all species, including our own.

Parasitism and infectious disease

Many animals and plants live in intimate association with other organisms of different species. In some cases, these associations are of mutual benefit to both organisms as, for instance, in the case of lichens. Each lichen consists of an organised network of filaments of a fungus and cells of algae are entangled in this network. The algae carry out photosynthesis and so contribute large energy-containing food molecules to the complex, while the fungus provides support and absorbs water and soluble nutrients from the environment.

The word symbiosis is used to describe mutually beneficial associations of this kind. Such associations may involve animals or plants, and they are sometimes obligatory, sometimes optional, sometimes permanent and sometimes transient. In the case of the lichens, the algae can grow independently, but the fungus cannot.

Parasitism is a type of association between two organisms in which one of them, the parasite, is dependent on and lives at the expense of the other, the host. The host provides the parasite with a habitat and with nourishment. Internal parasites live within the host's body, as in the case of the parasitic worms that are found in the intestines of animals, and various bacteria that live and multiply in internal organs. External parasites, like the fleas of mammals and the mistletoes of plants, live on the outside surface of the host, but still derive nourishment from it. Parasitism is extremely common, and there would not be a single species of multicellular animal or plant that does not normally harbour parasites of one kind or another. The parasites of mammals include not only many kinds of single-celled organisms, but also roundworms, tapeworms, hookworms, liver flukes, mange mites, lice, ticks and fleas.

While all parasites feed on nutrients supplied by their hosts, they can also cause varying kinds of damage. Under typical natural conditions animals are not seriously disadvantaged by the parasites they carry; but unnatural crowding or ill health from other causes often leads to damaging levels of parasitic infestation.

Many animal parasites (i.e. parasites that are animals) are highly host-specific, and can only establish themselves in, or on, the particular species to which they have become adapted through evolution. The tapeworm *Taenia saginata* cannot infect any species other than humans. Some other parasites can live in, or on, a wide range of host species. The adult form of *Trichinella spiralis*, the tiny worm that causes trichinosis in humans, can live in the small intestines of humans, pigs, walruses, rats, beavers, racoons, skunks, seals, bears, polar bears, wolves, lynx and many other mammals, as well as some birds.

Some animal parasites have complicated life cycles involving two or even three different species of hosts. The small tapeworm, *Echinococcus granulosus*, which lives in the intestines of members of the dog family, periodically sheds its final segment, which contains fertilised eggs, and these are excreted into the environment with the dog's faeces. If the eggs are then taken up and swallowed by sheep or cattle, they hatch in the animal's intestines, giving rise to very small 'hooked embryos'. These embryos burrow through the walls of the intestine into the bloodstream and eventually become lodged in the lungs or liver, or occasionally some other organ, where each embryo develops into a round, fluid-containing sac called a *hydatid cyst*. In each of these cysts large numbers, sometimes millions, of minute 'tapeworm heads' grow from the cyst lining. The cysts remain in this form until the animal dies. If the affected organ is then eaten by a dog, the cyst is broken, and the tapeworm heads become attached to the wall of the dog's intestine, to grow into adult tapeworms, so completing the life cycle. Hydatid cysts sometimes develop in humans who have swallowed the eggs of the tapeworm picked up from an infected dog.

While the vast majority of bacteria, protozoa and fungi are free-living and incapable of multiplying in the bodies of living animals and plants, some have become adapted through evolution to a parasitic way of life. In order to do so, they must acquire resistance to the host's natural defence processes, which normally detect and eliminate foreign cells.

Infection with parasitic microorganisms often causes signs of overt disease. Well-known examples in plants include potato blight and wheat rust, both of which are caused by a fungus. Examples in humans include malaria and dysentery, which are due to protozoa; and tuberculosis and cholera, which are due to bacteria. The mechanisms by which disease-causing microbes cause damage to the host's tissues are variable. In some infectious diseases the injury is due to toxic substances produced by the invading organism, as in the case of diphtheria and tetanus. In others, the inflammatory response of the host's tissues is a major cause of distress, as in the case of tuberculosis, plague and pneumonia.

Most disease-producing bacteria, protozoa and fungi are obligatory parasites and they are incapable of multiplying outside the bodies of their hosts. There are, however, a few microorganisms that are normally free-living but which can, under certain circumstances, multiply in animal tissues and cause disease. An example is the bacterium *Clostridium tetani*, which lives naturally in the soil. If this organism gains access to the body of an animal through a wound and becomes surrounded by dead tissue in a relatively oxygen-free environment, it may be able to multiply. When it does so, it produces a protein that is extremely toxic for most mammals, causing the symptoms of tetanus.

Some of the more severe infectious diseases of both plants and animals are caused not by bacteria, protozoa or fungi, but by viruses, most of which are not visible under the light microscope. They range in size from the virus of foot and mouth disease, which has a diameter of only 21 millimicrons, to cowpox virus, which measures 210 x 260 millimicrons. Most bacteria measure 1,000 to 2,000 millimicrons. Viruses are relatively simple structures, with a central core of nucleic acid that is usually surrounded by a layer of protein. They are only capable of multiplying within living cells.

The presence of a virus in the cells of a host does not necessarily cause any serious harm, and viruses can sometimes lie latent in the body tissues for long periods without giving rise to any symptoms. Some plant viruses can cause a mottling effect on the leaves or on the petals of the flowers, without apparently interfering significantly with the plant's viability. On the other hand, some viruses cause severe disease. In humans, infectious diseases due to viruses range from relatively mild conditions like the common cold and gastric flu, to herpes, influenza, measles, mumps, poliomyelitis and smallpox. In most virus diseases, the immune response is effective in bringing the infection to an end. In the more severe diseases, like poliomyelitis and smallpox, however, serious and lasting damage, and sometimes death, may come about before the immune response is effective.

When the tissues of a mammal or a bird are invaded by microbes of a kind that the animal has not experienced previously there usually occurs an immediate inflammatory response in which mobile cells known as phagocytes attempt to ingest and digest the intruding organisms. In cases when the microbes are able to withstand these mechanisms, a second phase of the defence process comes into play — the immune response. As a result, after about a week or 10 days, the animal's tissues become more sensitive to this particular microorganism and its products. This increased sensitivity is associated with the appearance in the blood and other body fluids of antibodies, which are protein molecules that have the property of combining specifically with the macromolecules that the invading microorganism produces. Because of the presence of the antibodies and some other specific changes in the host's tissues, the cellular reaction to the infectious agent is greatly enhanced and much more effective. In the case of cholera, for example, if infected humans live long enough for the immune response to develop properly, they are likely to survive and overcome the infection.

The specific immunity produced in this way is usually long-lasting, so that if the host becomes infected with the same kind of pathogenic organisms again at some time in the future, the initial response of the tissues is likely to be immediate, vigorous and effective. All artificial immunisation procedures are based on the principle of bringing about an immune response against disease-causing organisms, and so conferring protection against infection at a later date.

The health of ecosystems

The concept of health can be applied not only to individuals and populations, but also to ecosystems and, even, the biosphere as a whole.

A healthy ecosystem can be defined as one in which the rate of plant growth, or photosynthesis, is more or less constant from year to year. An unhealthy ecosystem is one in which the annual photosynthesis is progressively declining. Another characteristic of healthy ecosystems, at least on a regional level, is the maintenance of biodiversity. That is, a healthy ecosystem contains a wide range of interacting and interdependent species of plants, animals and microorganisms.

Box 2.1 Health needs of ecosystems

- Concentrations of greenhouse gases in the atmosphere (e.g. carbon dioxide, methane, nitrous oxide) at, or near, natural levels
- The absence of polluting gases or particles in the atmosphere that interfere with living processes (e.g. particulate hydrocarbons from combustion of diesel fuel and sulphur oxides)
- The absence of substances in the atmosphere (e.g. CFCs) that result in destruction of the ozone layer in the stratosphere that protects living organisms from ultraviolet radiation from the Sun
- The absence of chemical compounds in the soil and in oceans, lakes, rivers and streams that can interfere with the normal processes of life (e.g. POPs and heavy metals)
- No ionising radiation that can interfere with the normal processes of life and photosynthesis
- The maintenance of biodiversity in regional ecosystems (including aquatic ecosystems)
- Soil loss no greater than soil formation (i.e. no soil erosion)
- No increase in soil salinity
- The maintenance of the biological integrity of soils (i.e. maintaining a rich content of organic matter)
- Intact nutrient cycles in agricultural ecosystems over long periods of time (requiring return of nutrients to farmland)

Source: Stephen Boyden

In most terrestrial ecosystems, organisms living in the soil play an essential role in maintaining the health of the system. Bacteria and fungi, of which there can be up to five tonnes per hectare, play a crucial role in making nutrients available for plant growth. In some productive ecosystems there can be 60 or more earthworms per square metre, and these contribute to the health of the system by breaking down large pieces of organic matter into humus.

Natural events, such as bushfires and unusual droughts, can temporarily interfere with ecosystem health.

As we shall discuss in later chapters, human activities sometimes seriously interfere with ecosystem health. From our understanding of ecosystem function and of the impacts of natural events and of humans on ecosystems, we can put together a working list of the health needs of ecosystems (Box 2.1).

Animal behaviour

Every species of animal has characteristic behaviour patterns that are appropriate and advantageous in its natural habitat. Some are aimed at procuring food, avoiding predators and building shelters or nests. Others are social, involving interaction with other members of the same species, as in sexual activity, parental behaviour, status-determining behaviour, mutual grooming and simply staying together in a group.

Many animals are fiercely territorial and will vigorously defend their living space against intruders, especially intruders of the same species and sex. In poplar aphids, for example, two females may be locked in a kicking and shoving contest for hours, even days, as each of them attempts to take over a potential gall site at the base of a leaf. Territorial behaviour is common among birds and, in many species, the male selects a territory in which he and his mate will build a nest and raise their young. Occasionally, birds attack not only members of their own kind, but any intruders, regardless of the species to which they belong. Australian magpies are notorious for their attacks on humans, and unprepared people who enter their territories in springtime can receive nasty wounds to their ears or heads. Territorial behaviour is also seen in many kinds of fish, reptiles and mammals.

On the other hand, there are also many species of animal that show no signs of territoriality, permitting other members of the same species to come and go without interference. There is also tremendous variation in other aspects of social behaviour. Some animals spend most of their lives as part of a social group, such as a flock or a herd, while others live continually with just one partner of the opposite sex. Others spend most of their lives in solitude.

There is an important distinction between innate and learned behaviour. Innate behaviour is programmed by the genetic information that the animal inherits from its parents. Learned behaviour is the consequence of previous experience, and of learning ways of achieving pleasurable sensations or experiences, and of avoiding unpleasant ones. Many actions are mixtures of both innate and learned behaviour.

The contribution of learning to behaviour patterns becomes increasingly important in animals higher in the evolutionary scale, and in humans it plays a bigger role than in any other species. But even in humankind there are still some innate or genetically determined tendencies to behave in certain ways in certain situations. Apart from obvious innate behaviours, such as sucking in newborn infants and the tendencies to eat when hungry and to drink when thirsty, it is highly probable that the innate behavioural tendencies of our species lie behind much social behaviour. There is a strong case for the view that the tendency for people to identify with an in-group, to seek the approval of the members of this group and to avoid their disapproval is innate. Environmental factors, however, especially cultural pressures, largely determine the criteria of approval and disapproval.

An aspect of animal behaviour that has received much attention over the years can be summed up in Alfred Tennyson's famous phrase 'Nature, red in tooth and claw'. This notion became linked in people's minds with the phrase 'struggle for survival' that was in common use after the publication of Darwin's theory of evolution. Today there are many television documentaries that focus on the tearing apart and eventual consumption of one animal by another. Indeed, carnivorous animals routinely engage in this kind of behaviour to keep alive.

There is an important perspective, however, that is often ignored in discussions on this topic. The great majority of animals spend most of their lives in a state of good health and relative tranquility, perhaps enjoying themselves most of the time. The really nasty bit — being attacked and eaten by another animal — is only a fraction of their whole life experience. In any case, it might be preferable to a long, drawn-out and painful death from chronic disease. Moreover, in some mammalian species, the release of endorphins during attack by a predator may well significantly reduce the pain and distress experienced by the prey.

3

Humans enter the system

Evolutionary background

During the last part of the dinosaur era, around 65 million years ago, there existed on Earth a small group of tree-dwelling primates that looked something like present-day shrews. Among them were the ancestors of humankind.

By 5 or 6 million years ago in the African savannah, there were some much larger primates walking with an upright posture. One particularly well-preserved fossil is that of a young female, which was found in Ethiopia and dated to about 3 million years ago. She is known informally as Lucy, and the species she belonged to has been called *Australopithecus afarensis*. She had a skull very like that of a chimpanzee, with a brain of around 500 cubic centimetres.

Two and a half million years ago there were primates in Africa making stone tools. One species, called *Homo habilis*, was only 90–120 centimetres tall, but it had a brain with a volume of about 800 cubic centimetres, which is about 300 cubic centimetres larger than that of a chimpanzee. These animals consumed both plant and animal food. Another, rather similar species was *Homo rudolfensis*, and it existed at about the same time.

Around 1.8 million years ago, a taller species named *Homo ergaster* was living in the same area. This was the first hominid to have an essentially modern form, and it is possible that the three species *Homo habilis*, *Homo rudolfensis and Homo ergaster* were all living in the same region at the same time.

It seems that *Homo ergaster* spread out of Africa into Europe and Asia about 1 million years ago, giving rise eventually to a form of humanity referred to as *Homo erectus*. Remains of this species have been found as far east as Java, Indonesia. These people were 150 to 180 centimetres tall and their skulls had an uninterrupted bar of bone above the eyes. Their brains were between 1,000 and 1,200 cubic centimetres. *Homo erectus* may have survived in eastern Asia until around 300,000 years ago, and possibly even to 25,000 years ago.

Fossil remains have been found in Europe and Africa of humans that lived 300,000 to 800,000 years ago and that appear to be intermediate between *Homo erectus* and modern humankind, although their relationship to earlier and later forms is unclear.

From about 200,000 years ago, and during most of the first part of the fourth, or Würm, glaciation, western Europe was occupied by a distinctive form of humanity classified as *Homo neanderthalensis*. These people were of stocky build and most of the men were a little over 152 centimetres tall, and the women a little shorter. Their skulls were flattish on top and noticeably rounded at the back, and they had the pronounced brow ridge reminiscent of *Homo erectus*. They had massive musculature and jaws and the brains of adults ranged from 1,450 cubic centimetres to 1,650 cubic centimetres in volume, which is rather larger on average than that of modern humankind. They were well acquainted with the use of fire, hunted big game and dressed in animal skins. They used paints to decorate their bodies and sometimes they buried their dead. Recent evidence suggests that there were other human species or subspecies living in Asia at around this time.

It is now clear that humans with the physical characteristics of our own species, *Homo sapiens*, were in existence in Africa and probably elsewhere, around 200,000 years ago. They were tall people with rounded skulls and steep foreheads, and their average cranial capacity was about 1,400 cubic centimetres. They had well developed chins, and their brow ridges were only moderately developed and were not continuous from side to side. If we could bring some of them back to life, dress them in modern clothing and set them loose on a city street, they would be indistinguishable from some of the better specimens of modern humanity.

The emergence of human culture

Over the hundreds of thousands of years that these anatomical changes were taking place in our hominid ancestors, something else was also happening of tremendous significance. This was the evolutionary development of the ability of humans to invent and memorise symbolic language, and to use it to communicate among themselves. This linguistic aptitude depends both on characteristics of the human brain and on special anatomical arrangements in the larynx, pharynx and tongue, which permit us to utter an amazing range of different sounds.

Along with the evolution of the aptitude for symbolic language, there was a parallel emergence of the capacity to compose, make and enjoy music. As in the case of human language, this trait is unique in the animal kingdom.[1] The making and appreciation of music has become a hugely important aspect of human experience, and this has possibly been so since the very earliest days of our species.

The aptitude for language led to the progressive accumulation of shared knowledge, beliefs and attitudes in human groups. That is, it led to human culture.

Another characteristic of human behaviour is the ability to invent new technologies and to pass on this technical know-how from one individual to another and from generation to generation. Some other primates and some birds exhibit a trace of this ability. In humans, the aptitude for technology is greatly assisted by the extraordinary dexterity of our species and by spoken or written language.

The rapidity of the development in evolution of the human capacity for language and culture indicates that, once a rudimentary ability to invent and use symbolic language emerged, it was immediately of major biological advantage. Under the prevailing conditions its chief

---

1   Some birds, of course, sing; but the pattern of sound is largely genetically determined.

advantage probably lay in its role in the exchange and storage of useful information about the environment. The fact that culture was of biological advantage under the conditions in which it evolved does not mean, however, that it will necessarily still be advantageous under conditions that differ significantly from those of the evolutionary environment.

Human culture thus came into existence as a new kind of force in the biosphere — a force destined eventually to bring about profound and far-reaching changes across the whole planet. Its impacts on the rest of the living world during the long hunter–gatherer period of human history were, however, modest in comparison with those of later times after the advent of farming, and especially after the Industrial Revolution.

Geographical distribution and genetic and cultural variation

Despite their unimpressive physical strength, biting power and speed of running, *Homo sapiens* proved to be biologically successful. They spread from Africa across Asia and, by 60,000 years ago, they had reached Australia. Around 45,000 years ago they became the dominant human type in Europe and Asia, displacing the Neanderthals and other human species or subspecies. They were responsible for a marked diversification and sophistication of culture, as reflected in the many kinds of artefacts they left behind in the form of scraping tools, knives, burins, awls, needles, spatulas, weapons of various kinds, pendants, necklaces, armbands, musical instruments, statuettes and rock paintings.

By 40,000 years ago, modern humans had spread throughout Africa, Asia, Europe and Australia. It seems that they did not arrive in the Americas before 35,000 years ago, and perhaps not until around 16,000 to 13,000 years ago.

The spread of humankind across the globe was associated with some divergence in the genetic characteristics of populations, resulting in observable differences between people living in different parts of the world — differences, for example, in stature, facial features and colour of skin, hair and eyes. Other genetic differences, like variability in the distribution of blood-group antigens between populations from different regions, are detectable only by scientific procedures. These differences were brought about partly by chance and partly by natural selection. It has been suggested that a light skin is of selective advantage in areas of the world where sunlight is relatively weak, because it allows the formation of vitamin D below the skin. In tropical areas a dark skin would prevent the synthesis of excessive amounts of the vitamin, which could be harmful, and would protect skin cells against damaging radiation from the Sun.

There was also, of course, much cultural divergence. Human groups across the world developed countless different religious belief systems and they eventually spoke in thousands of different languages.

Humans in their natural environment

The term 'natural environment' is used here to mean the environment in which a species evolved and to which it has become genetically adapted through natural selection. The natural environment of the human species is, therefore, the environment of our hunter–gatherer ancestors. There have been too few generations since those days for there to have evolved major genetic changes rendering humans significantly better adapted to the conditions of modern civilisation. Biologically, we are basically the same animal as our forebears of, say, 15,000 years ago. We share their innate biological characteristics.

General biology

The conditions of life of hunter–gatherers varied significantly from time to time and place to place. They lived in many different kinds of habitats, ranging from dense tropical rainforests, semitropical savannah and deserts, through to temperate forests and grasslands to the ice-covered plateaus in the far north of Europe, Asia and America. Most of them, however, spent their lives in mild to warm, relatively fertile areas with moderate rainfall.

The following brief account of the typical conditions of life and ecology of primeval people covers mainly those aspects that are likely to have been universal or, at least, usual among hunter–gatherers.[2]

As in the case of animals living under natural conditions, most of the time most of the people had plenty of food and were well nourished. The typical diet consisted of a wide range of different foods of plant origin, including berries and other fruits, nuts, roots, grains and leaves, and a certain amount of cooked lean meat making up roughly 20 per cent of the diet by weight. The meat had a low fat content and a high ratio of polyunsaturated to saturated fat. The diet of newborn infants was, without exception, human milk.

Diets varied, however, from one region to another. Eskimos consumed a higher proportion of meat than people living in the African savannah where, at some times of year, the diet contained only a small amount of animal protein.

The time spent collecting food varied according to local circumstances but, in general, averaged two to three hours a day. Much of the food was brought back to the group's campsite for sharing.

An essential activity associated with the acquisition of food was the manufacture of weapons for hunting and of tools for cutting meat and scraping animal skins. Stone spearheads, axes and most other weapons and implements were usually made by men, while ornaments were made by both men and women.

The amount of food energy required by humans depends on both their size and their pattern of behaviour. In our own society, an adult male leading a rather sedentary life might use about 10 megajoules per day. About half of this energy is used in basic metabolic processes and the rest in voluntary muscular activity. The same person, leading a moderately active life, would use about 12 megajoules a day, but he could use as much as 30 megajoules if he was performing exceptionally heavy work in a cold climate.

---

2   The term 'primeval people' is used here to mean people who are hunter–gatherers. 'Primeval society' means hunter–gatherer society.

It is likely that the daily use of food energy by adult males in hunter–gatherer communities would have been between 15 and 20 megajoules. Some idea of the relationship between food energy and physical work can be gleaned from the following facts: one teaspoonful of sugar (about 0.1 megajoules) is sufficient fuel for an adult male to run for five minutes; and it takes seven hours of non-stop moderate cycling to 'burn off' 0.5 kilograms of body fat.

Patterns of rest and sleep in primeval society varied according to circumstances. In general, people tended to sleep or to rest when they felt like it, and when there was nothing better to do. Most sleep was taken during the hours of darkness, but short naps were also common during the daytime.

Reliable figures are not available on fertility rates in recent hunter–gatherer populations. It seems, however, that a common picture was for couples to have three or four children, two or three of which could be expected to reach adulthood and to become parents themselves.

Box 3.1 A generational perspective

Picture yourself on the stage of a large theatre with room for an audience of 2,000. In your mind's eye, place your mother in the seat at one end of the front row, and then her mother next to her and so on — until you have filled the place with 2,000 generations of mothers and daughters.

The great majority of your maternal ancestors in the theatre would have known nothing of agriculture or the urban way of life. Only the women in the front 20 or so rows would have been alive since the time when farming first began, and only those in the front six or seven rows would have lived after the earliest cities came into existence, although very few of them are likely to have actually lived in cities.

You could fill at least one other similar theatre with earlier maternal hunter–gatherer ancestors belonging to the species *Homo sapiens*. All these women really existed, and they lived in a state of health, at least until the birth of a child.

If you were to carry out the same mental exercise using an amphitheatre with room for 200,000 people, the individuals in the rows at the back would not be members the genus *Homo*.

Source: Stephen Boyden

Social organisation and psychosocial aspects of life experience

Humans are social animals and, in hunter–gatherer societies, each individual belonged to a close-knit group or band in which there was constant exchange of information on matters of mutual interest. The great majority of social interactions were between individuals who knew each other well. The size of these bands was variable and was largely determined by prevailing ecological conditions. People were very aware of their responsibilities as determined by the prevailing social norms, and there was a good deal of coming and going between neighbouring bands.

There was a certain division of labour within bands in that most of the gathering of plant foods and small animals was done by women, while the hunting of larger animals was mainly a male activity. Women played the major role in caring for small children.

Other activities included various forms of creative behaviour, including making tools, weapons and ornaments, and telling stories. Much time was spent in conversation and making music and dancing were common activities.

Serious physical violence and other antisocial behaviour within bands was probably not common, at least not under reasonably favourable ecological conditions. Arguments among group members certainly occurred, however, sometimes leading to physical violence. The frequency of such violence varied from group to group and from time to time.

The behaviour of humans in the natural environment, or indeed any other environment, can be described at two levels: basic behaviour and specific behaviour.

Humans share certain basic behavioural tendencies. These range from the obvious: behaviours that are linked closely with physiological functions, to behaviours of a psychosocial nature, which are more difficult to define. The former include the tendencies to eat when hungry, to drink when thirsty, to copulate when appropriately stimulated, and to move from less comfortable to more comfortable positions. The more psychosocially oriented basic behavioural tendencies include the tendencies to identify with an in-group, to seek the approval and avoid the disapproval of members of the in-group, to show loyalty to members of the in-group, and to seek to maintain or improve status within the in-group.

While these basic behavioural tendencies are evident in all human populations, their specific manifestations vary greatly according to circumstances and, especially, according to the prevailing culture. While humans all over the world tend to seek the approval of the members of the in-group with which they identify, the criteria of approval, which are largely culturally determined, differ from one group to another. This particular basic behavioural tendency can, in different cultural settings, lead to specific behaviours as different as baking a cake or throwing a bomb into a crowded bus.

Culture can also reinforce or suppress basic behavioural tendencies. In some societies, the tendency to compete is greatly reinforced while, in others, it is suppressed. Creativity is encouraged in some cultures, but discouraged in others.

The topic of human aggression has received a great deal of attention from academics. Some have argued that there is an innate aggressiveness in humankind towards other humans, especially in the case of males. According to one school of thought, this aggressiveness builds up in the individual until it finds a behavioural outlet. Others argue that it is culture that determines whether humans behave aggressively towards one another. The view taken here is that, while humans are not 'innately aggressive', there is an innate tendency for people to behave aggressively in response to perceived threats to themselves or to their in-group. Cultural factors have an important influence on what is, and what is not, perceived as a threat. Aggressive behaviour can also be a consequence of cultural pressures in societies in which aggression and violence are seen as criteria for praise and approval.

Judging from the evidence derived from research into contemporary hunter–gatherer groups, primeval societies were not characterised by constant hostilities with neighbouring bands. On the other hand, violent interaction, sometimes resulting in deaths, certainly occurred from time to time.

While overt physical aggression towards out-groups is not inevitable in humankind, there does seem to be a universal tendency of humans to be suspicious of strangers.

There is universal concern among humans for the well-being of members of the in-group to which they belong, and especially of children, but there is no evidence for an innate concern for the well-being of members of out-groups.

In most hunter–gatherer societies there was no rigid hierarchical structure. Leadership was usually determined spontaneously and was based mainly on prowess and personality. One individual might be leader in a hunt, another in a honey-collecting expedition, another in music-making or dancing, and another in religious rituals.

Apart from the division of labour based on age and gender, there was no occupational specialisation of the kind found in later societies. All the women took part in collecting plant foods, sometimes assisted by men, and all the men participated in hunting and making weapons.

Other psychosocial features of the primeval lifestyle include the following:

- most individuals were part of a care-giving and care-receiving network
- there was considerable variety in daily experience
- the immediate environment was full of interest to everybody
- most people experienced personal creative behaviour on a daily basis
- there was spontaneity in behaviour
- aspirations were short term and of a kind likely to be fulfilled
- most people experienced a sense of personal involvement in daily activities, a sense of belonging to a group and to a place, a sense of challenge and a sense of responsibility.

Children

The life conditions of children in primeval societies reflected the spontaneous nature and relatively uncomplicated social organisation of the communities. Babies were kept close to their mothers for the first year or so of life but, after this, they might be left at the camp to be minded by relatives or friends when their mothers went gathering. For the first few years, children were indulged by their parents and other members of the band and they were seldom severely punished for transgression of norms. Customs differed from one hunter–gatherer society to another with regard to the control of behaviour in children after the age of five or six, although it is likely that the laissez-faire attitude persisted in most groups.

The learning experience of children did not involve any formal program of teaching. The process was spontaneous and was based on such basic behavioural characteristics as the tendencies to imitate and to seek approval. Much learning in childhood took the form of listening to, observing and copying slightly older children, who were in turn learning from their older siblings or peers. Playing games based on mimicry of adult behaviour was universal among hunter–gatherer children. Mild aggressive behaviour was not uncommon in children, although it seldom resulted in serious injury.

Arts and crafts

The human species is noted for its capacity to make and use tools. Some other species of animals, however, also use tools on occasion. Otters make use of stones to break open molluscs, and both chimpanzees and some birds not only use sticks of wood for a variety of purposes but sometimes also manufacture simple implements; and some birds use short sticks to prise insects from under the bark of tree trunks. Humans, however, are clearly more adept at inventing, manufacturing and using tools than any other species.

In primeval society, most individuals spent part of each day making or shaping something by hand or creating new patterns in some other way. Recent cultural developments have reduced the incentives and opportunities for creative behaviour for a high proportion of the population, depriving people of an important source of enjoyment and self-fulfilment. Much has been written about human creativity, but most of this literature has been concerned with the activities only of quite exceptional specimens of humanity, such as da Vinci, Michelangelo, Picasso, Mozart and Beethoven. Less attention has been paid to creative behaviour as a feature of the everyday life of ordinary people.

Ecology

While the existence of language and the use of hunting weapons certainly had some effects on the ecological interrelationships between humans and other species living in their immediate environment, the overall ecological impact of hunter–gatherers was not great. They fitted into the food chains in much the same way as any other omnivorous species — that is, they acquired their energy in the form of organic molecules in plant and animal foods, and they were eventually consumed by predators, scavengers or decomposing micro-organisms. In the ecosystems in which hunter–gatherers lived, it is likely that only about 1/10,000 to 1/100,000 of the energy fixed by photosynthesis actually flowed through the human population.

The capacity for culture, together with human dexterity, did lead to one particularly important difference between the ecology of our species and that of other mammals. The regular use of fire and the manufacture and use of tools added an extra dimension to the metabolism of human populations — referred to as technometabolism.

Technometabolism is defined as the pattern of flow of energy and materials into, through and out of a human population that results from technological processes. It contrasts with biometabolism, which is the flow of energy and materials into, through and out of human organisms. Technometabolism is a product of cultural evolution and is a new phenomenon in the history of life on Earth. It is of tremendous significance ecologically and in many other ways.

The use of fire, in particular, was a development of enormous ecological significance. It was the first example of the regular and deliberate use by humans of extrasomatic energy — that is, energy used outside the human body, as distinct from the somatic energy, which is consumed in food and which flows through the human body.

It has been estimated that the introduction of the regular use of fire in human populations approximately doubled the per capita energy use, bringing the total energy used per day per person (men, women and children) to about 16 megajoules — that is, roughly 8 megajoules used in biometabolism and 8 megajoules in the burning of wood.

Health

As in the case of all other animals living in their natural environments, most of the time most of the members of hunter–gatherer communities were in a state of good health. Indeed, they had to be in order to survive and successfully reproduce under the demanding conditions of their lifestyle and habitat.

Most people:

- would have been well nourished. There is no diet better for any animal than that which is typical of its natural lifestyle and environment. Undernutrition, malnutrition and obesity were rare in normal circumstances, although, in periods of unusual drought, people's health would have deteriorated

- would not, before contact with people from urban societies, have suffered from such infectious diseases as influenza, the common cold, measles, cholera, typhoid and plague. There were simply not enough humans living together to support these pathogenic microbes

- would not have suffered from such organic disorders as appendicitis, duodenal ulcer, diverticular disease of the colon and cardiovascular disease. It is known that blood pressure tends to remain more or less constant in adults in primeval societies after the age of about 20 years, rather than increasing steadily after this age, as is frequently the case in modern communities.

On the other hand, the primeval lifestyle was characterised by some built-in hazards that are absent from modern society. There was a considerable risk of serious injury acquired during hunting, and severe wounds often became infected, leading to gangrene or septicaemia. Any incapacitation due to injury or ill health was of much greater survival disadvantage in the hunter–gatherer setting than under the protective conditions of modern civilisation. People did not have the benefit of the artificial antidotal measures like antibiotics, chemotherapeutic agents and surgery that are available today. The average age in primeval populations was typically around 25 years.

Comment on human health needs

Numerous definitions of health have been proposed for humankind. Here we adopt a biological definition similar to that given above for all living organisms: health in humans is that physical and mental state that would have been likely to ensure survival and successful reproduction. Another appropriate definition is: health is that state of body and mind conducive to, and associated with, full enjoyment of life.

Of greater practical interest than the definition of human health is the identification of the health needs of our species. There are various approaches to this issue, ranging from one's personal experience to the application of knowledge from medical research. Here we adopt a biological line of attack based on the evolutionary health principle (see the section 'Health and disease' in Chapter 2), taking the life conditions of humans in their natural environment as a starting point. These are the conditions to which our species was genetically adapted through evolution and they satisfied the survival, health and reproductive needs of our ancestors for many thousands of generations. In accord with the evolutionary health principle, significant deviations from these conditions are likely to be associated with signs of maladjustment or ill health.[3]

The evolutionary health principle clearly applies to a wide range of physical aspects of life conditions in humans. There is no diet better for humankind than that which was typical for hunter–gatherers. It is also clear that the principle is applicable to some aspects of behaviour. Marked deviations from natural sleeping patterns cause

---

3   This does not mean that evolutionary change in the human species has come to a halt. There has been a relaxation of some selection pressures that were powerful in the hunter–gatherer environment and, in the long term, this will result in genetic changes in human populations (J.M. Rendel, 1970. 'The time scale of genetic change'. In S. Boyden (ed.), *The impact of civilisation on the biology of man*. Australian National University Press, Canberra). There have also been some new selection pressures associated with the advent of farming that have produced changes in some populations. A well-known example of this is the emergence and spread in European populations of lactase production into adulthood in response to the availability of bovine milk as a food source. For discussion of this change and for other examples, see G. Cochran & H. Harpending, 2009. *The 10,000 year explosion: How civilisation accelerated human evolution* (Basic Books, New York).

maladjustment, and health is likely to be impaired if patterns of physical exercise deviate markedly from those of humans in the natural habitat.

There are good reasons for supposing that the principle also applies to psychosocial and relatively intangible aspects of life experience. The conditions of life of hunter–gatherers are usually characterised, for example, by incentives and opportunities for creative behaviour, a sense of personal involvement in daily activities and plenty of convivial social interaction. Most of us would agree that such conditions are likely to promote health and well-being in our own society.

Taking our knowledge of the conditions of life of hunter–gatherers as a starting point, we can put together a working list of the physical and psychosocial conditions that are likely to promote health and well-being in our species (Box 3.2). They are referred to as universal health needs because they apply to all members of the human species wherever or whenever they may be living.

Most of the items on the list of postulated psychosocial health needs, like creative behaviour and sense of personal involvement, cannot be defined and measured as easily as physical health needs, but this does not mean that they are less important.

It is true that not every item on this psychosocial list is absolutely essential for health. Lack of satisfaction of one health need may be offset to some extent by the satisfaction of others. On the other hand, every item on the list will, if satisfied, make a positive contribution to health and well-being.

Unfortunately, conventional measures of social well-being in our society today do not take the less tangible psychosocial aspects of life experience into account, nor do they feature on the platforms of the major political parties. It is important, however, that deliberate effort be made to take these less tangible factors into consideration in any assessment of current human life conditions or in planning for the future.

Box 3.2 Universal health needs of the human species

Physical needs

- clean air (not contaminated with hydrocarbons, sulphur oxides, lead, etc.)
- a natural diet (i.e. calorie intake neither less than nor in excess of metabolic requirements; foods providing the full range of the nutritional requirements of the human organism, as provided, for example, by a diverse range of different foods of plant origin and some cooked lean meat; a diet that is balanced in the sense that it does not contain an excess of any particular kind of chemical constituent or class of food; foods with the physical consistency of natural foods and containing fibre; foodstuffs devoid of potentially noxious contaminants or additives)
- clean water (free of contamination with chemicals or pathogenic microorganisms)
- absence of harmful levels of electromagnetic radiation (e.g. alpha, beta, gamma, ultraviolet and x-rays)
- minimal contact with parasites and pathogens
- protection from extremes of climate (temperature, wetness)
- noise levels within the natural range
- a pattern of physical exercise that involves some short periods of vigorous muscular work and longer periods of medium (and varied) muscular work.

Psychosocial needs

- an emotional support network, providing a framework for care-giving and care-receiving behaviour, and for exchange of information on matters of mutual interest and concern
- the experience of conviviality
- opportunities and incentives for cooperative small-group interaction
- levels of sensory stimulation that are neither much lower, nor much higher, than those of the natural habitat
- opportunities and incentives for creative behaviour
- opportunities and incentives for learning and practising manual skills
- opportunities and incentives for active involvement in recreational activities
- opportunities for spontaneity in human behaviour
- variety in daily experience
- short goal-achievement cycles and aspirations of a kind that are likely to be fulfilled
- an environment and lifestyle that is conducive to a sense of personal involvement, purpose, belonging, responsibility, challenge, comradeship and love
- an environment and lifestyle that do not promote a sense of alienation, anomie, deprivation, boredom, loneliness, or chronic frustration.

Source: Stephen Boyden

Brief comment is called for on the concept of stressors and meliors. Stressors are experiences that cause anxiety and distress and they are a normal aspect of life. If they are short-lived and not too severe, they can be seen as contributing positively to the quality of life and well-being; but if they are excessive and if they persist, they can interfere seriously with both mental and physical health. Equally important are experiences that have the opposite effect to stressors and that give rise to a sense of enjoyment. Such experiences have been called meliors. Common meliors include the experience of creativity, fun, aesthetic enjoyment and conviviality.

Every person can be considered at any given time to be at some point on a hypothetical continuum between a state of distress and a state of enjoyment. Their position on this continuum is largely a function of the balance between meliors and stressors in their recent experience.

The cultural environment has an immense influence both on the levels of meliors and stressors in an individual's daily experience as well as on the nature of the factors that cause them. Culture also affects the energy and pollution costs of attempts to avoid stressors or to experience meliors.

4

The Early Farming and Early Urban phases

Ecological Phase 2: The Early Farming Phase

Modern humans were in existence for around 200,000 years before some of them, in several different parts of the world, decided to take up farming.

It is likely that the shift from an economy based on hunting and gathering to one in which human populations were mainly dependent on farming for their food was a slow process that extended over 1,000 years or more. This deliberate manipulation of the processes of nature for human ends resulted in big changes in the interrelationships between people and the other living organisms in the ecosystems of which they were a part. Eventually, it also led to big changes in the interrelationships between humans themselves.

People started farming independently in at least three regions of the world: south-western Asia, south-eastern Asia and middle America. The earliest evidence of agriculture based on the cultivation of seeds, notably wheat and barley, and the domestication of goats and sheep dates back to about 12,000 years ago. It is found in high country in a broad area known as the Fertile Crescent, extending eastwards from Greece to a region to the south of the Caspian Sea.

Farming eventually spread northward from the Middle East and, by 6,000 to 7,000 years ago, it had reached southern Germany and Holland, where people were growing einkorn, emmer, barley, peas, beans and flax, and keeping goats, sheep, cattle and pigs. Meanwhile, other farming people were also moving through Spain and Switzerland to France and eventually Britain. By 4,500 years ago, they had reached the southern edges of the vast coniferous forests of Scandinavia and Russia.

The domestication of plants for food production was well established in Thailand by 9,000 years ago. Instead of cultivating wheat and barley, these farmers grew various roots, like taro and yams, as well as cucumbers, peppers, bottle gourds and beans of different kinds. They also systematically and deliberately encouraged the growth of banana, coconut, sago and breadfruit trees, and they domesticated pigs and poultry.

It is likely that maize had been domesticated in the Americas by 8,000 or 7,000 years ago and, by 5,000 years ago, squash, beans, avocados, gourds, pumpkins and chillies were also cultivated. Potatoes were grown in Peru 7,000 years ago.

Health and disease

The changed conditions of life associated with farming had important effects on the relationships between human populations and potential parasites and disease-producing microorganisms. Microbial and parasitic diseases became more common. It has been suggested that it was the introduction of slash-and-burn agriculture to tropical and subtropical regions that allowed malaria to become an important cause of ill health and death, because it brought humans into close contact with populations of mosquitoes breeding in stagnant pools of water. This situation, combined with the relatively high density and immobility of the population, was particularly favourable for the maintenance of the malarial parasite's life cycle.

Schistosomiasis occurs everywhere in the world where large-scale irrigation is practised, and there is good evidence that it was already common in Mesopotamia, Egypt and the Indus Valley several thousand years BC. This disease is due to infection with one of several species of blood fluke that spends part of its life cycle in the body of certain species of snail. It has been estimated that about 200 million people in 72 countries suffer from schistosomiasis today.

The increase in mortality due to various parasitic diseases did not, however, outweigh the biological advantages of village life. By the time that the early Mesopotamian cities came into being, there were probably between 50 and 100 million people on the Earth.

The ecological impacts of early farming

Although farming took on many different forms in different places, all early food-producing activities consisted in essence of the deliberate redistribution of plant and animal species in a given area, aimed at increasing the local concentrations of edible species and at reducing concentrations of species of little or no food value. As a result, instead of covering big distances in search of food, people spent most of their time in and around one place, protecting and propagating the desired plants and animals and keeping out unwanted species. Special techniques were developed to enhance plant productivity in the different regions in which people settled, such as hoeing, irrigation and, eventually, ploughing and manuring.

The end result of this change in subsistence behaviour was that a given area of land yielded much more food for humans than had been the case in hunter–gatherer ecosystems. A larger proportion of the energy fixed by photosynthesis in the local area was available, as somatic energy, to pass into and flow through human beings. Population densities of hunter–gatherer communities ranged from around 0.02 to 0.2 persons per square kilometre and, in early farming societies, they ranged from 25 to 1,000 persons per square kilometre. It has been claimed that peasant farming in southern China supported 7,500 persons per square kilometre.

Generally speaking, the amount of food produced per hour of human effort in early farming societies was probably not very different from that acquired by hunter–gatherers in the same length of time.

One of the ecological consequences of farming was the dispersal of animal and plant species, not just locally in and around the farm, but eventually across the whole world. The invasion of the Americas by Europeans was an especially important development in this context, resulting in the introduction into Europe of maize, potatoes, squash, gourds, pumpkins, peppers, chillies and turkeys. It also caused movement in the opposite direction, with the introduction into the American continent of the cereal crops of Europe, and of cattle, sheep and pigs. Later, all the main European food sources of the northern hemisphere were introduced to New Zealand and Australia, where they now form by far the greater part of people's diet.

To be successful in the long term, farming practices had to be ecologically sustainable, which is to say they had to maintain the bioproductivity of the food-producing ecosystems indefinitely. In many regions farming practices were sustainable in this sense, although this often required temporarily abandoning, or resting, a food-producing area to allow it to regain fertility, as in the case of slash-and-burn agriculture and the two- and three-field systems adopted in Europe. This rotation was necessary because farming involved the continual removal from a plot of land of nutrients, which were incorporated in edible plants or animals. These nutrients were ultimately deposited somewhere else, in the form of excrement, corpses or other organic waste material, thereby disrupting the natural nutrient cycles. Such resting was not necessary in rice paddies, because incoming water continually brought in a new supply of nutrients.

In some regions, however, farming led to progressive loss of productivity, either due to soil loss through erosion, as in parts of northern Africa, or because of salinisation, as in the irrigation farming ecosystems in Mesopotamia.

Diseases of plant or animal food sources due to parasites or disease-causing microbes were also a frequent cause of serious, but temporary, food shortages. One reason for this is the fact that high population densities, or monocultures, of any particular species of plant or animal are especially susceptible to infectious or parasitic diseases. Important examples affecting plant foods have been wheat rust, ergot (a disease of rye) and potato blight. This last was responsible for the Irish famine of 1845 and 1846, which caused hundreds of thousands of deaths and widespread human distress. It also had long-lasting political repercussions.

Ecological Phase 3: The Early Urban Phase

By around 9,000 years ago, some farms in the south-western corner of Asia and the south-eastern corner of Europe were producing more food than was necessary to satisfy the nutritional needs of the farming communities. The existence of this surplus made it possible for fairly large clusters of people, sometimes consisting of several thousand individuals, to aggregate together in townships, with many of them no longer playing any part in food production.

An extraordinarily interesting example of one of these early townships is Çatal Hüyük in Anatolia, where a thriving community of 5,000 or more people lived between 8,000 and 9,000 years ago. Their food sources were primarily barley, wheat, peas, lentils, sheep, goats and, later in the period, cattle.

Twelve building levels have been uncovered at Çatal Hüyük, spanning a period of at least 1,000 years. There is no sign of any invasion or sudden cultural change during this period, and accidental fires may have been responsible for the rebuilding that occurred. The mud-brick buildings were built close together and, in each case, the entrance was through an opening in the roof. There were few lanes or passages, but plenty of courtyards, which were apparently used as depositories for rubbish.

By 5,500 years ago, cities had formed in the valley of the Tigris and Euphrates rivers in Mesopotamia, and some of them had populations of 50,000 to 60,000 people. There is evidence that thriving cities and trading centres with pyramids, temples and ordinary houses existed 5,000 years ago in Peru.

Most of the early cities in Sumer, Mesopotamia, had populations of 10,000–20,000, possibly reaching 50,000 in the case of Uruk. Athens probably had a population of about 100,000 in the days of Pericles and Socrates.

Until the 17th and 18th centuries AD, the populations of cities in Europe and the Middle East did not, with a few exceptions, exceed 100,000. The exceptions were Rome, with a population variously estimated at between 350,000 and 1 million at about the time of Jesus Christ and, about 300 years later, Constantinople, with a population of about 1 million.

During the 17th century there were big increases in the populations of some cities and, by the end of the 18th century, Paris had a population of over 670,000, Naples of over 430,000 and London of over 800,000.

From the biological standpoint, three fundamental changes lie at the root of all other factors that set cities apart from earlier human societies, and all of them had far-reaching biological and cultural repercussions.

First, there was the spectacular increase in population density, which meant that there was an enormous increase in the number of other members of the human species encountered by the average person in the course of his or her daily activities. Unlike the case in farming and primeval societies, many of those encountered were not members of the individual's personal in-group.

Second, most of the people living in cities were not farmers. For the first time in human history, a significant proportion of the population went through life without participating in the intimate interactions with the natural environment associated with the food quest. For sustenance they relied on the activities of others. This was not only a novel development in the history of the human species, it was unique among the vertebrates.

The third outstandingly important development accompanying urbanisation was the shift from relative homogeneity to heterogeneity in human populations. This change was made up of three basic interrelated elements: occupational specialism, political stratification and wealth stratification. Material wealth was itself a new factor in human experience.

Division of labour, or occupational specialism, has been described as the hallmark of civilisation. Already in the cities of Mesopotamia 5,000 years ago there were leather workers, cabinet-makers, potters, metalworkers, basket-makers, shopkeepers, brewers, stonecutters, weavers, gardeners, artists, music-makers, soldiers, scribes, priests, and other officials in the administration and members of the ruling families. Relatively permanent social hierarchies came into existence, ranging from kingship, priesthood and the aristocracy through to commoners and slaves.

Health and disease

The conditions of life of early urban dwellers deviated even more than those of farmers from the conditions to which our species was adapted through evolution. As would be expected from the evolutionary health principle (Chapter 2), this led to further new patterns of ill health and mortality. Moreover, because there were big differences in the conditions of life in different sections of the population, different groups of people experienced different patterns of health and disease.

It is an epidemiological principle that, when animal populations become significantly more dense than in the evolutionary environment of the species, disturbances occur in host–parasite relationships. Overcrowded conditions provide greatly increased opportunities for the spread of parasites and pathogenic microbes.

The concentration of large numbers of people in urban settings greatly facilitated the spread of potentially pathogenic organisms in various ways. One of the most important of these was spread of infection by touch — either when infected persons touch non-infected persons or when infected persons touch objects, such as tools and doorhandles, which are then touched by non-infected persons.

Many pathogenic organisms, such as those responsible for typhoid, cholera, infantile diarrhoea and dysentery, are spread by direct or indirect contact with human urine or faeces. In many early cities, accumulating masses of human excrement were a constant menace to health, especially when they contaminated water supplies.

Another reason why the increase in the size and density of human populations resulted in greater prevalence of infectious diseases is the fact that many pathogenic microorganisms require a large contiguous population in order to exist. It has been calculated that the measles virus needs a contiguous population of 300,000 people to keep it going. If the measles virus had come into existence in primeval times and had infected a member of a hunter–gatherer band, it would have passed on to other members of the group, some of whom might have died, while others would have recovered and then been immune; and so the measles virus itself would have died out. This would also be true of the several hundred relatively mild acute virus infectious diseases that circulate in human populations today. They are all products of civilisation.

So it came about that microbial disease became firmly established as a feature of urban living, and it played a major role in human affairs well into the 20th century. Sometimes it caused terrible epidemics but some diseases, such as tuberculosis and infantile diarrhoea, were constantly present in human populations. Infectious disease was by far the most important cause of death in the early urban societies, and it remained so until the processes of cultural reform came into play in the second part of the 19th century. Apart from smallpox, typhus and the plague, the most important of the infectious diseases during the Early Urban Phase were dysentery, enteric fever, typhoid, tuberculosis and, late in the period, cholera.

Turning to nutrition, the tendency for some urban populations to rely on one or two staple foods resulted in specific deficiency diseases, like rickets, scurvy, pellagra, beri-beri, and vitamin A deficiency. Urban populations were also often affected by famine, leading to under-nutrition and starvation.

In summary, while urban conditions provided protection from some of the important causes of death in hunter–gatherer and early farming societies, such as attack by predators and schistosomiasis, the new diseases of early civilisation caused death rates in cities to be very high — probably considerably higher than in rural areas.

The populations of these urban societies were probably maintained by continuing immigration from rural areas, where by far the greater part of the total human population lived. The end result was an overall increase in the total human population from around 5 to 10 million when farming was first introduced to about 600 million in 1700 AD. This is a doubling time, on average, of around 1,500 to 2,000 years.

The world population just before the beginning of the present Exponential Phase of human history, around 200 years ago, was about 1 billion.

Ecological impacts of early cities

In the early days of towns and cities, their main ecological impact was probably deforestation, due to the demand for timber for the construction of buildings and for burning to provide heat for cooking, comfort and various industrial activities. In the Middle East, the great forests of cedars of Lebanon disappeared early on, as did the evergreen forests of southern Europe. Deforestation occurred progressively throughout the rest of Europe, and also around urban areas in Middle America.

Other sources of extrasomatic energy were watermills and windmills. Wind power also played a role as the driving force for sailing ships.

Taking society as a whole, however, the use of extrasomatic energy per person did not increase dramatically in the Early Farming and Early Urban phases and, on average, probably seldom exceeded 30 megajoules per day.

Culture

The topics of conversation among the inhabitants of the early cities were very different from those of hunter–gatherers or farmers, reflecting the new occupational structure of society, the elaboration of religion and people's lack of involvement in the acquisition of food.

One of the most significant cultural developments in the first part of the Early Urban Phase was the introduction of writing. It has been suggested that very simple forms of writing to record information may go back to around 11,000 years ago. But it was in the early cities that it became important for keeping economic records, such as tallies of cattle and sheep, measures of grain and jars of butter. It also came to be used for royal inscriptions, legal codes, religious texts and recording legends and political events.

The biohistorical importance of writing lies in the fact that it meant that information could be stored outside human brains. The storage capacity of the brain was no longer a limiting factor, so that society came to amass vastly more information than had been possible previously; and writing enabled people to communicate about things in some detail without having to see each other.

The concept of ownership of physical property, which was only weakly developed in hunter–gatherer communities, became much stronger in urban society. Eventually it came to apply not only to land, animals and material objects, but also to other members of the human species. In the early Mesopotamian cities, slaves were plentiful, and slavery persisted in the Western world until the lifetime of the grandparents of some of us alive today. It still exists in one form or another in many parts of the world today.

A significant economic development was the introduction and increasing importance of a cash economy. The desire to exchange things is as old as humankind, but it is not older. No other species engages in exchange of material objects, although giving and taking is common among animals. When such exchange took place in hunter–gatherer and early farming societies, it usually involved simple barter. In early urban societies and in some early farming communities, however, a different system was introduced involving a new factor, namely money, which eventually became an all-important element in human society. Money is essentially an arrangement in which the prevailing culture bestows a certain agreed symbolic value on certain objects that can then be used as a medium for the exchange of goods and services among individuals and groups. It has taken many different forms, including sea shells, coins of silver, gold or other metal and paper notes, all of which can be given in exchange for goods or services. Today the modern equivalent of money sometimes exists only in the memories of computers.

Another important development in the early cities was the introduction of formalised codes aimed at controlling, or preventing, various forms of aggressive behaviour. Punishments, ranging from monetary fines to death, were imposed for transgression of these laws. This was a cultural adaptive response to the fact that many different groups of humans were now living in close proximity to each other, creating a potentially inflammatory situation because there is no innate tendency in humans to avoid aggressive behaviour between out-groups.

Perhaps the most significant of all cultural changes associated with urbanism resulted from the fact that large numbers of people were separated from the natural environment and played no role in the acquisition of food, so that they no longer felt part of nature — and urban cultures evolved that regarded the natural world as alien and threatening.

Cultural maladaptations

There are countless instances of cultural maladaptation in the Early Farming and Early Urban phases of human history. A particularly tragic example was the ancient Chinese custom of foot binding, which prevented the normal growth of the feet of young girls and caused them excruciating pain. This extraordinary practice well illustrates the propensity of culture to influence people's mindsets in ways that result in activities that are not only nonsensical in the extreme, but also sometimes cruel, destructive and contrary to nature. This particular cultural maladaptation was mutely accepted by the mass of the Chinese population for some 40 or more generations. Such is the brainwashing power of culture.

Only a few generations ago, Western culture regarded slavery as entirely appropriate and British imperialism was completely acceptable in the lifetimes of the parents of some of us alive today.

Throughout the history of civilisation, different cultures, including our own, have come up with a fascinating range of delusions about how social well-being, or prosperity, can best be achieved, and some of these have led to patent examples of cultural maladaptation. According to the dominant culture of the Mayan civilisation, prosperity could best be achieved by pleasing the gods, and the best way to please the gods was to torture, mutilate and then sacrifice human beings. This behaviour can be regarded as a cultural maladaptation because it was the cause of a great deal of unnecessary human suffering, and it clearly did not do the Mayans any good. Their civilisation collapsed suddenly, perhaps for ecological reasons, around 900 AD.

Again the point to be emphasised is the fact that, while there may well have been a handful of sceptics among the Mayans, the great majority of them believed that the torture and sacrifice of humans was an entirely appropriate behaviour.

Another example is provided by the Jukun people of Nigeria in the 19th century. At that time their culture included the belief that their king, who was elected for a period of seven years, was a kind of 'living reservoir' of the various forces that caused the soil to be fertile and the seeds to flourish, and that generally brought health and well-being to the people. The king therefore had to be protected from any kind of injury because, if he became sick or lost blood, some of the beneficial forces might escape, with undesirable consequences for the population. Indeed, if he fell off a horse, became seriously ill or impotent, he was quickly strangled and a new king was elected. In any case, the king was put to death when the seven years of his reign were up, strangulation being used so that no blood would be spilled. This particular instance is perhaps not a cultural maladaptation, in that it does not seem to have done any real harm, except from the point of view of the king himself at the end of his reign.

Cultural gullibility is indeed a fundamental characteristic of our species.

5

Ecological Phase 4:
The Exponential Phase

Introduction

Towards the end of the 17th century and during the 18th century, the intellectual movement commonly referred to as the Enlightenment was underway in Europe. This movement emphasised rational thought, as opposed to religious tradition, as a means of understanding the universe and improving the human condition.

This fourth cultural watershed paved the way for the fourth, Exponential Phase of human history — a phase that is associated with further profound changes in the ecological relationships between human populations and the rest of the biosphere. Especially significant ecologically were the introduction of machines that used extrasomatic energy from fossil fuels for performing various kinds of work and the enormous growth of the chemical industry. Also of great importance was the discovery and application of electricity, radioactivity and radio waves.

There have also been important scientific advances that have resulted in a massive growth in the human population. There are now about 1,500 times as many people alive as there were when farming began. Nearly 90 per cent of this increase has occurred in the Exponential

Phase of human history. This vast increase in the number of people on Earth is putting immense pressures on the food-producing ecosystems of our planet.

Cultural maladaption

Cultural maladaptations in ecological Phase 4 are manifold. Some affect humans directly, while others cause damage to the living environment on which we depend. At present, some even pose a threat to the survival of civilisation, perhaps of the human species

Over the past 150 years, new knowledge coming from the sciences has often led to warnings about the undesirability of maladaptive activities, setting in motion cultural responses aimed at overcoming the cultural maladaptations. This process is referred to as cultural reform.

Cultural reform is complicated and involves prolonged interactions between different interest groups in society. A key role is often played initially by minority groups, occasionally by single individuals, who start the ball rolling by drawing attention to an unsatisfactory state of affairs. A good example is Rachel Carson who, in her groundbreaking 1962 book *Silent Spring*, drew attention to the insidious and destructive ecological impacts of certain synthetic pesticides.

Almost invariably, these expressions of concern coming from reformers are promptly contradicted by others, the counter-reformers, who set out to block the reform process. This predictable backlash often involves, but is not restricted to, representatives of vested interests who believe that the proposed reforms will be to their disadvantage.[1] They are likely to argue that the problem does not exist or that it has been has been grossly exaggerated, and they try to ridicule the reformers by calling them alarmists, fanatics, scaremongers and prophets of doom. Nowadays some of the counter-reform forces are extraordinarily powerful.

---

1   For a detailed discussion in the context of tobacco smoking, CFCs and climate change, see N. Oreskes & E.M. Conway, 2010. *Merchants of doubt.* Bloomsbury Press, New York.

Eventually, if the reformers are successful, a change comes about in the dominant culture and members of government bureaucracies and other organisations set about working out ways and means of achieving the necessary changes. Their efforts may still be hindered by the stalling tactics of counter-reformers.

Cultural reform may be corrective or antidotal. Corrective reform occurs when the adaptive process involves correcting the underlying cause of maladjustment or disharmony. An example is provided by the restoration of vitamin C to the diet of a population suffering from scurvy. In antidotal reform, the unsatisfactory conditions that are the underlying cause of disturbance are not modified, and the adaptive response is aimed at alleviating the symptoms or at an intermediate factor. Most, but not all, of the work of the medical profession is antidotal rather than corrective

Some examples of serious cultural maladaptation in the Exponential Phase will be discussed below.

Life experience

In the year 1900, around 20 per cent of the world population lived in cities. By 1989 the proportion had grown to 40 per cent. Today it is greater than 50 per cent and about half of all urban dwellers live in cities with populations of 100,000–500,000. Less than 10 per cent live in cities with populations of more than 10 million.

One of the notable differences between the lifestyles of modern city dwellers and those of their hunter–gatherer ancestors is the clear distinction in the present setting between work and non-work activities. Linked with this is the fact that there has been a progressive trend for people to become further and further removed from the end product of their efforts. Individuals employed as computer operators may be playing an essential role in outcomes as diverse as manufacturing ammunition and protecting biodiversity.

Other significant changes in human behaviour include:

- spectacular increases in the speed and distance of human travel
- widespread instantaneous electronic communication between humans across the world

- most adults now perform much less physical work than the typical hunter–gatherer (although there are some exceptions, such as marathon runners and some farm workers)
- for many people, much of their physical activity is unnaturally repetitive (e.g. computer operators, professional violinists)
- most people now spend several thousand hours a year sitting still and staring at the rectangular screen of a television set or computer.

Another important psychosocial change is the lengthening of goal-achievement cycles. In the natural environment most goals were set for a few hours — or at most a few days. In present society goals are often set for many years ahead.

An all-pervasive feature of affluent societies in the modern world is consumerism, which seems to make a definite contribution to human well-being. For many individuals, the act of purchasing manufactured goods is an important source of enjoyment, tending to counter the undesirable effects of various environmental stressors. A common and often effective response to a feeling of depression is a shopping spree. Consumerism has perhaps come to replace, or compensate for, more biosphere-friendly forms of enjoyment that were important in earlier societies, such as the experience of conviviality and various kinds of creative behaviour and activities that resulted in a sense of personal involvement and purpose.

Differences in material wealth are extreme and are increasing in many nations, including the United States, Australia, the United Kingdom, India and China. The 125 richest people in the world possess assets greater than all the least developed countries combined. In Australia, the income of senior executives is 150 times average weekly earnings; and the seven richest Australians hold more wealth than 1.73 million households in the bottom 20 per cent range.

Human health

Infectious disease

The public health movement in Britain, which began in the early 1800s, is an early and well-documented example of cultural reform. A small group of reformers, consisting mainly of young medical doctors who were well aware of the appalling life conditions of the working class in the new industrial towns, called for profound improvements in urban sanitation and housing.

The efforts of these reformers were promptly countered by counter-reformers in the form of rich landlords and representatives of water companies, whose financial interests might have been threatened by government action aimed at alleviating the situation. While this backlash slowed down the reform process, major and effective reforms eventually came into place, beginning with the *Public Health Act* of 1948.

Since that time there has been a spectacular drop in the incidence of serious contagious diseases and malnutrition, especially in developed countries. The following factors have been especially important:

- *greatly improved sanitation*, which reduced the likelihood of contact with disease-producing organisms spread via human excreta
- *improved nutrition*, leading to a greater resistance to infection
- *artificial immunisation*. This was first introduced in Britain in the late 18th century in the case of smallpox, but is now applied to a wide range of infectious agents. It has resulted in a significant drop in the incidence of many infectious diseases. Indeed, it led to the actual elimination of smallpox in 1979. Rinderpest, a serious virus disease of cattle and other ungulates in Africa has also been eradicated as the result of an effective vaccination campaign
- *the introduction of antibiotics*, which has further reduced mortality from bacterial infections.

At the present time, some disease-causing bacteria, including 'golden staphs' and bacteria responsible for tuberculosis and pneumonia, are developing resistance to previously effective antibiotics. This is becoming a very serious problem.

There is still room for big improvements in the developing world. It is estimated that over 1 billion people do not have access to safe drinking water. Adequate sanitation is not available to 2.5 billion people.

Although there has been an impressive drop in the incidence of serious bacterial disease in the developed countries, modern populations the world over are exposed to an ever-increasing number of viruses that cause relatively mild diseases, like the common cold, influenza and various gastro-intestinal disturbances. These viruses need a large contiguous human population to keep them going, and they could not have survived before the advent of towns and cities. But current conditions suit them well and, as new viruses arise, there is nothing to stop them circulating around the global population of humankind ad infinitum. There are now hundreds of these viruses in existence.

Some scientists believe there is a strong likelihood that further new viruses with high mortality rates will emerge in the future, spreading rapidly through the global population.

Nutrition

As a result of improved scientific understanding of the nutritional requirements of the human species, urban populations are now much better nourished than was the case a couple of hundred years ago.

Significant deviations from the natural, or evolutionary, diet are still, however, the cause of much unnecessary ill health. Overeating is a major problem. According to the World Health Organisation, in 2013 obesity had doubled worldwide since 1980. In 2008, 35 per cent of adults over the age of 20 were overweight, and 11 per cent were obese. In 2011, 40 million children under five years old were overweight. In Australia 60 per cent of adults and one in four children are now overweight. Excess body weight is known to contribute to cardiovascular disease, diabetes, osteoarthritis and some kinds of cancer.

Other deviations from the natural diet that contribute to ill health include the consumption of refined carbohydrates and the absence of sufficient plant fibre in the diet.

Incidental contamination of food with the chemical products of industrialisation has proved a serious problem over the past half century. Contamination with pesticides, such as DDT and its breakdown product DDE, has been especially significant. Another common contaminant is polychlorinated biphenyl (PCB), which is used for various industrial purposes.

In the developed countries there has been growing awareness of the risks associated with chemical pollution of foods, leading to government regulations aimed at controlling the use of potentially toxic substances.

Turning to deliberate additives, countless different chemical substances are put into human food for various reasons. The most widely used of these is sodium chloride (salt), which humans have added to their food since ancient times. The majority of people in our society consume 10 to 15 times more salt than is necessary to satisfy their physiological requirements. A strong body of medical opinion holds that this deviation from the natural diet is responsible for much of the high blood pressure that is common in modern societies.

In 1820, Friedrich Accum, a German chemist living in London, published a *Treatise on adulteration of food* in which he denounced the use of chemical additives in food. This was a groundbreaking work and his book sold well. One of his concerns was the practice of adding alum to bread to make it look whiter. There was a vicious backlash from counter-reformers, however, in the form of bakers and millers, and Accum received many threats to his life. He eventually left London to return to his homeland.

Many years later, the validity of Accum's claims was confirmed and legislation was introduced aimed at preventing adulteration of bread with any officially unapproved substance.

Today, commercially prepared foods contain a wide range of additives with specific functions, such as preservatives, anti-oxidants, colouring agents, flavouring agents, sweeteners, filling agents, stabilisers, emulsifiers and other 'improving agents'. The battery of flavouring agents in today's commercial food products includes well over a thousand different chemical compounds.

The advances in nutritional science over the past hundred years, including the discovery of vitamins and essential amino-acids, represent one of the most impressive chapters in the annals of scientific research, and they have contributed immensely to human health and well-being. On the other hand, it is sobering to bear in mind that no knowledge of the existence, chemistry or biological function of vitamins or any other nutrient is necessary for the avoidance of nutritional deficiency diseases. All that is required is, first, understanding of the evolutionary health principle (Chapter 2); and second, knowledge that the typical diet of *Homo sapiens* in the natural habitat of the species consisted of a wide variety of different kinds of fresh vegetables, fruits, nuts and roots, and some cooked lean meat.

Alcohol consumption

No doubt some of our hunter–gatherer ancestors had the occasional meal of fermented fruit leading to a pleasurable feeling of mild intoxication. Once humans adopted the farming lifestyle, however, they lost no time in learning how to brew alcoholic drinks. It seems that, at least by 9,000 years ago, grapes, berries, honey and rice were being used to produced alcoholic beverages in northern China, while people in the Middle East were making barley beer and grape wine.

Wine drinking was a feature of ancient Greece, where people who did not drink wine were considered barbarians. Alcohol consumption was also common in ancient Egypt and Rome. But it is notable that in all these cultures, drunkenness was deplored, except in the case of certain religious festivals. Moderation was the law of the day.

Before the Middle Ages, the main alcoholic beverages in Europe were beer and wine. Distillation leading to the production of spirits like gin, vodka and whisky became widespread in the 15th century.

The Christian Church in Europe in the 16th to 19th centuries considered alcoholic drinks to be a gift from God — to be enjoyed for pleasure and health reasons — but again in moderation. Drunkenness was seen as a sin.

Less is known about the history of alcohol production and consumption in the Americas, although a wide range of alcoholic beverages from different plants were known to the indigenous inhabitants before contact with Europeans.

Today there is big variation among different countries in alcohol consumption. For instance, the per capita consumption of pure alcohol in Austria, Ireland and France is about 12 litres per year, while in Mexico it is 5.3 litres; Israel, 2.4 litres; and Turkey, 1.6 litres. In Australia, the United Kingdom, and Switzerland it is about 10 litres.

The effects of alcohol consumption on the human organism are well known. On the positive side, it gives rise to an enjoyable state of mind and it facilitates pleasurable social interaction. In many situations, the consumption of alcohol can be seen as an adaptive response to the requirement of modern society that individuals interact with a large number of complete strangers. Under the influence of alcohol, natural reserve and suspicion give way to an atmosphere of relaxed conviviality. Also on the positive side, if the statistics can be believed, is the fact that people who drink a moderate amount of alcohol regularly are likely to live longer than those who abstain.

On the negative side, the consequences of heavy drinking include lack of coordination, blurred vision, interference with judgement and sometimes, but not inevitably, aggressive behaviour. It is estimated that 50 per cent of road accidents in Australia are due to the overconsumption of alcohol. Excessive alcohol consumption has a range of other undesirable consequences. These include loss of jobs, family disruption, memory loss and various physiological disorders like cirrhosis of the liver. It also increases the likelihood of some forms of cancer.

Tobacco smoking

The story of tobacco smoking provides a good example of cultural maladaptation, cultural reform and cultural counter-reform. Right from the early days of smoking in Europe, occasional individuals intuitively felt this unnatural behaviour was not a good thing. One such person was King James I of England (James VI of Scotland). In 1604 he described tobacco smoking in the following words:

> A custom loathsome to the eye, hateful to the nose, harmful to the brain, dangerous to the lungs, and in the black, stinking fume thereof, nearest resembling the horrible Stygian smoke of the pit that is bottomless.[2]

---

2   James I of England, VI of Scotland, 1604. 'A counterblast to tobacco'. *Oxford Dictionary of quotations*. 2nd edn. 1956. p. 256.

In fact, from the 16th century onwards, smoking was banned in many Catholic churches throughout Europe and in Mexico. Smoking was banned in several European cities, including Berlin, in the 18th and 19th centuries. In the early 1940s, the Nazi regime mounted antismoking campaigns and attempted to restrict smoking in government offices, universities and some hospitals.

The reform movement was boosted by the epidemiological studies of Richard Doll and his colleagues in the early 1950s, which showed without doubt that tobacco smoking is the cause of a great deal of ill health and early mortality. The predictable counter-reform backlash from vested interests was, however, still active some 20 years after this work. I have in my possession a pamphlet from that time that was distributed by the Australian cigarette industry as 'an information service to smokers'. On the front page there is a single quotation as follows:

> The concept that smoking is the cause of the increase in lung cancer and emphysema is a colossal blunder.

Inside the pamphlet there are a few more quotations by members of the medical profession, taken from a public enquiry into smoking and health before the 1969 Committee on Interstate and Foreign Commerce, United States House of Representatives. Another quotation reads as follows:

> A bandwagon effect has resulted even in the medical and scientific community where too many have accepted the pronouncements (against smoking) of dedicated zealots, lacking the time to examine the scientific basis, or lack of it, for such pronouncements.

The following quotation from the 1972 annual report of Philip Morris (Australia) Limited provides another example:

> During the year, the Commonwealth and several State governments passed new laws relating to the sale of cigarettes. In addition, the Commonwealth Government intends to provide funds to a Commonwealth–State committee to finance a three-year anti-smoking campaign. The new Commonwealth law requires the addition of a government health notice at the conclusion of all broadcast commercials. The regulations introduced by the States to [sic] prohibit the sale by retailers of cigarettes if the packaging does not carry specified printed

health notices. These new restraints derive from the widely published assertion, which is unsupported by valid experimental or clinical evidence, that cigarette smoking is a direct cause of certain illnesses.

Your Directors, in common with the cigarette industry around the world, have continuously evaluated all the evidence which has been put forward from time to time in an attempt to support this assertion. They remain convinced that the case sought to be made against smoking is not proved; that there are major and obvious faults in the arguments and statistics put forward by the proponents of the anti-smoking thesis.

The above extract was reported in an editorial in the *Medical Journal of Australia* on 3 February 1973. In the words of the editorial:

These bland statements are made in flat (almost dead-pan) contradiction of the enormous mass of carefully assessed and freely available evidence of the harmful effects of cigarette smoking on health. Much of it is consolidated in the two reports of the Royal College of Physicians of London, the 1971 and 1972 reports of the United States Surgeon-General, and the reports of the First (1967) and Second (1971) World Conferences on Smoking and Health.

It seems that the cigarette companies now recognise that they have lost this battle. Evidence of this is provided by a more recent report from Philip Morris to the Czech Republic where this firm has about 80 per cent of the cigarette market. According to the *Guardian Weekly*: 'The report found that the economy received a number of benefits from smoking. There was income from excise duty and health cost savings due to early mortality.'[3] It was pointed out that, in 1999, the Czech Government saved up to the equivalent of US$32 million on health care and pensions for the elderly, thanks to premature deaths from cigarettes. The total benefits outweighed the costs of health care for sick smokers and loss of income tax from deceased workers. The net benefit to the government from the smoking population was calculated at US$143.5 million.

---

3   *Guardian Weekly*, 19–25 Jul. 2001, pp. 221–22.

The first decade and a half of the 21st century has seen the introduction of regulations banning smoking in enclosed spaces and public places in many countries across the world, led by Bhutan and Ireland. In Bhutan it is illegal to sell tobacco. Recently, cities in China, including Beijing, have introduced restrictions on smoking in public places.

There has been a significant decline in the number of people who smoke in most Western countries. In Australia, the proportion of men and women who smoked in 1976 was 43 per cent and 33 per cent respectively. In 2014–15, it was 16.9 per cent and 12.1 per cent.

Although the psychological effect of nicotine is less dramatic than drugs like heroin and cocaine, its addictive power is just as great, if not greater. Typical withdrawal symptoms include a craving for nicotine, headaches, irritability, anxiety, sleep disturbances, hunger, difficulty concentrating and a lowered heart rate and blood pressure. Most symptoms peak in the first day or two and then lessen, although the craving for a cigarette may persist for months.

In the United Kingdom, two-thirds of smokers want to give up, and about half of these try to do so every year. Only about 5 per cent of these are not smoking a year later. In Australia, 73 per cent of smokers have tried to give it up.

Educational programs about the undesirable effects of tobacco smoking on health, therefore, seem to be only minimally effective in persuading smokers to stop. Most smokers are well aware of the damage their habit is likely to do to their health.

One approach is to try to combat the habit by administering nicotine in the form of chewing gum, lozenges or skin patches, which deliver a dose that is big enough to overcome the craving, but too small to produce a high. Nicotine alone is far less dangerous than tobacco smoke and it is much less carcinogenic, but it has some undesirable effects on the cardiovascular system. Nicotine replacement therapy is intended as a temporary measure to lessen the withdrawal symptoms, and so increasing the chances of quitting. However, 90 per cent of people who try nicotine replacement therapy take up smoking again within 12 months.

Another approach is suggested by the experience in Sweden, where 17 per cent of men are smokers. Another 19 per cent, however, 'suck tobacco' in the form of a product called 'snus', which is moist ground tobacco that is placed between the tongue and the lip, either loose or contained in a small permeable bag. The nicotine quickly reaches the blood stream, giving rise to the pleasurable high. Snus can, therefore, be regarded as a recreational drug, and the men who use it are not trying to break their addiction to nicotine. Swedish men have by far the lowest risk of dying from smoking-related diseases in Europe — 11 per cent, compared with 25 per cent for Europe as a whole.

Ecology

The most striking ecological characteristic of the Exponential Phase of human history has been the spectacular increase in the overall scale and intensity of human activities on Earth and their impacts on the living systems of the biosphere.

Population

Advances in the medical and nutritional sciences in ecological Phase 4 have resulted in a massive increase in the human population.

At the time when farming was introduced, the total human population probably stood at around 5 million and, 12,000 years later, in 1810, it had grown to 1 billion. It took only 120 years (1810–1930) for the next billion to be added. The next billion took 30 years (1930–60); the next billion, 15 years (1960–75); the next billion, 14 years (1975–89); and the next billion, 10 years (1989–99). The next billion took slightly longer, 12 years (1999–2011), bringing the total human population to 7 billion.

Throughout much of human history, life expectancy at birth has been around 20 or 30 years. During ecological Phase 4 it has increased remarkably, and the global figure in 2013 was about 67 years. It is considerably higher than this in the developed countries. In Australia, for example, life expectancy at birth in 2013 was 79 years for males and 84 years for females.

The current global population is increasing at the rate of around 1.4 million every week.

Food sources

The industrial transition led to a new era in farming involving fundamental changes in agricultural practice, especially in cereal production. From the ecological standpoint, one of the most significant aspects of the modernisation of agricultural systems has been the change in the energetics of crop production. In the early farming societies, the energy content of food was around 15 times as much as the energy spent in collecting it. In the United States today, the energy content of food received in the home is only about one-fifth of the energy used in farming practices, transportation, preparation and wrapping. At least 90 per cent of the energy input is in the form of fossil fuels.

Another feature of the transition has been an ever-increasing application of artificial fertilisers. Phosphate and potash fertilisers are derived from natural deposits, and world resources of phosphate rock are considered to be sufficient to last for 100 to 200 years, while potash reserves may be sufficient to last about 5,000 years. Nitrogen fertilisers are now made synthetically from atmospheric nitrogen. Although there is unlikely to be any shortage of nitrogen, the methods used are energy-costly, and this may create problems in the future. The overall global use of artificial fertilisers has increased about five-fold since 1950.

Other changes have included the widespread use of synthetic pesticides to control parasites and plant diseases and the cultivation of new, high-yielding varieties of certain crops.

All in all, these changes have resulted in big increases in crop yield per unit area.

Another striking development has been the great increase in yield per hour of human labour. In the Early Farming Phase a typical farming couple produced, like hunter–gatherers, sufficient food for themselves and their families and sometimes a small surplus to contribute to the diet of non-farmers. The situation is very different in the developed regions today. In the United States in the 1970s, one farm worker produced sufficient food for 50 people. Australia is an extreme case and, in a good year, one farmer now produces enough food for 85 people, two-thirds of whom live overseas. Many farmers today, however, work for 10 or more hours each day.

The modern farmer is dependent on the work of countless other individuals involved in the design and manufacture of tractors and other machinery, in the extraction and preparation of artificial fertilisers and in transportation of materials to and from the farm.

The average person in a typical industrial society today consumes, directly and indirectly, four-fifths of a tonne of cereal grain per year; but only about 10 per cent of this is eaten in the form of grain. Most of it goes to feed animals. A small proportion of this energy eventually reaches humans as meat, eggs, milk and cheese. As a result, overall less than 20 per cent of the food energy in the grain reaches humans.

The farming systems of the exponential societies also produce a broad range of vegetables and fruits for human consumption.

In the developing regions of the world, which include most of the rice-growing areas, farming has remained labour-intensive. Nevertheless, considerable increases in yield have been achieved in some regions as a result of the so-called Green Revolution. This movement began in Mexico the 1940s and spread worldwide during the 1950s and 1960s, and it continued to have an important influence on agricultural trends during the 1970s. The Green Revolution was based on the development of high-response varieties (HRV) of wheat and rice and, to some extent, of maize and millet, as well as the expansion of irrigation and the increasing use of artificial fertilisers and pesticides.

Unfortunately, the relief provided by the Green Revolution to food shortages in developing countries, where human populations are still growing rapidly, can only be temporary. These agricultural systems have reached their limits of production. In many regions, one social outcome of the introduction of the HRV varieties has been for the rich landlords to become richer and for the landless peasants to become poorer.

Growing appreciation that widespread use of fertilisers and pesticides can eventually have undesirable ecological and health consequences has led to the organic food movement. This movement dates back to the 1940s, but it did not really take off until around 1990. Organic foods are produced with farming methods that do not use chemical fertilisers or synthetic pesticides, although organic pesticides are used. Industrial solvents and chemical food additives are not used in the processing of organic foods.

The worldwide market for organic foods has grown rapidly since 2000. Many countries are establishing formal, government-regulated certification of organic food.

Scientific advances in genetics and molecular biology have recently led to the rapid development of the new field of genetic engineering, involving the artificial incorporation of genetic material (DNA) from one form of life into the genetic apparatus of another. This approach has an enormous number of applications, including the creation of food plants that are resistant to insect pests or to certain herbicides. In the latter case, the herbicides can be used freely for controlling weeds without fear of damaging the crop in question. Genetic engineering has already resulted in the creation of countless biologically novel forms of plants, bacteria, viruses and, more recently, animals. Most genetic modification of foods has, however, focused on crops in high demand such as soybean, corn, canola, and cotton seed oil.

There are strong differences of opinion about the wisdom and morality of genetic engineering and the consumption of genetically modified foods. One author writes: 'agricultural biotechnology is going to be one of the great disasters of corporate capital history',[4] while another takes the view that genetic engineering will, through its potential to increase yields, provide 'a much needed boost in the struggle to feed the world's growing population'.[5]

Land degradation

The production of food for over 7 billion people is putting immense pressures on the food-producing ecosystems of planet Earth.

Between 1961 and 2009, agricultural production expanded 150 per cent, due mainly to a significant increase in the yields of major crops. In many places, however, the better yields have been associated with land degradation, the main causes of which are soil erosion, loss of organic matter in the soil, disruption of natural nutrient cycles, soil salinity and various combinations of these forms of soil ill health.

4   See *New Scientist*, 6 Jun. 1998, p. 3.
5   J. Rifkin, 1998. *New Scientist*, 31 Oct. 1998, p. 34.

The United Nations' Food and Agricultural Organization (FAO) warns that the world's agricultural systems face the risk of progressive breakdown of their productive capacity as a result of excessive population pressure and unsatisfactory farming practices. According to the FAO, a quarter of agricultural land is already highly degraded. Another 8 per cent is moderately degraded and 36 per cent is classed as stable or slightly degraded; 10 per cent is described as 'improving'. The worst affected areas are along the west coast of the Americas, across the Mediterranean region of southern Europe and North Africa, the Sahel and the Horn of Africa, and throughout Asia. It is feared that these agricultural systems may not be able to satisfy human demands by 2050.

Climate change is expected to have a major effect on the world's food production, although there is much uncertainty about the extent and nature of these effects.

Water

Seventy per cent of water used by humans globally is used for agriculture. The proportion is much higher in many developing regions, and lower in the affluent countries where much water is used in industrial processes.

Severe water shortages now affect at least 700 million people across the globe and the number is expected to increase with predicted climate change. According to one estimate, more than half the world's population will be facing water shortage by 2025.

Loss of biodiversity

The present rate of extinction of living organisms is exceptionally high, due to the activities of humankind. According to one estimate, species are becoming extinct a thousand times faster than was the case in the late Pleistocene period (126,000–11,700 years ago), when the extinction rate was well above the average for geological time as a whole.

Some authorities believe that about a quarter or more of living organisms on Earth, excluding bacteria, archaea and viruses, will be extinct by 2025. Nearly 9,000 trees, representing about 10 per cent of tree species known to science, are threatened with extinction. Over 34,000 species of plants are on the verge of extinction.

The main cause of this loss of biodiversity is habitat destruction through various activities of humankind, including farming, logging, fishing and the construction of roads and buildings.

Deforestation

Progressive deforestation has been taking place in many parts of the world for centuries. Around 900 AD, 80 per cent of central Europe was covered with forest, but only 25 per cent in 1900. These forests reached their minimum at about the time of the First World War.

Globally, deforestation has increased especially rapidly over the past 60–70 years. Today, far more trees are destroyed every year than are planted.

Most of North Africa and the Middle East, as well as a great deal of continental Asia, Central America and the Andean regions of South America are now virtually treeless. Problems associated with deforestation are also looming large in parts of Central Africa and on the Indian subcontinent.

More than half of the world's tropical rainforests have been destroyed and the rate of destruction is accelerating. It has been estimated that tropical forest, which accounts for about 70 per cent of forest productivity in the world, is being destroyed at an annual rate of 4.1 million hectares in South America, 2.2 million hectares in Asia and 1.3 million hectares in Africa, giving a total of 7.4 million hectares per year. Some authors believe that the rate of decline is much greater than this. More than 20 per cent of the Amazon rainforest has already been destroyed. In 2004, the worst year on record, 27,000 square kilometres of this forest were destroyed and, recently, trees were being lost at the rate of 2,000 a minute.

If deforestation continues at current rates, scientists estimate nearly 80–90 per cent of tropical rainforest ecosystems will be destroyed by the year 2020.

Deforestation makes a major contribution to climate change. This is partly because healthy forests absorb a significant amount of the carbon dioxide emissions coming from human civilisation. The destruction of forests also results in the release of carbon into the atmosphere. Carbon emitted from this source now accounts for 12–17 per cent of all emissions resulting from human activities. This is more carbon dioxide than is given off by all cars, trucks, planes, trains and ships across the world.

Furthermore, deforestation is resulting in big losses in biodiversity, it is interfering with the water cycle and it is a major cause of soil erosion.

Industrialisation and technometabolism

The Exponential Phase of human existence has seen a massive surge in the intensity of technometabolism in human populations, with far-reaching ecological consequences. The overall pattern is depicted in Figure 5.1.

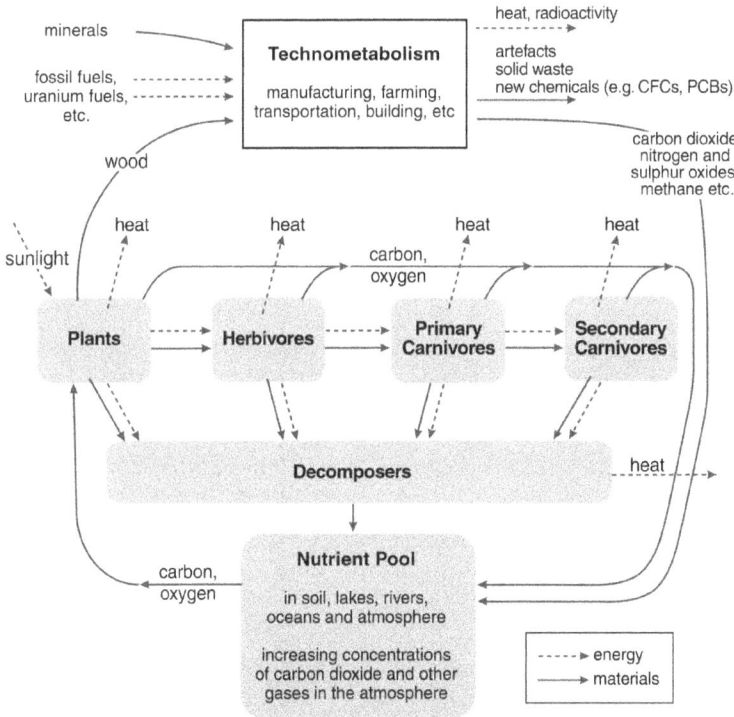

Figure 5.1 Flows of materials and energy in the modern world
Source: Stephen Boyden

Technometabolic inputs

Energy use is an important measure for a number of reasons. The rate of use of energy is probably the best single indicator of the overall intensity of human activity on the planet, since everything that we do involves a throughput of energy, although its impact will, of course, depend a great deal on the particular use to which the energy is put. It will also depend on the source of the energy, since some energy sources have by-products that have impacts on biological systems. These by-products include carbon dioxide and the oxides of sulphur and nitrogen from fossil fuels as well as the radioactive by-products from nuclear power plants.

The human species as a whole is now using about 20,000 times as much energy every day as was the case when farming began (Box 5.1). This is equivalent to the difference in weight between a small apple and a couple of tonnes of bricks. Well over 90 per cent of this increase has occurred in the past 100 years (Figure 5.2).

Box 5.1 Faster! Faster! Faster!: Energy use by humankind

The following analogy brings home the massive scale and recent intensification of human activities on Earth:

Let us suppose that farming began 12 hours ago (rather than 12,000 years ago) and that, at that time, humans jumped into a vehicle they had invented. The speed of this vehicle is proportional to the total amount of energy used each day by humankind. Energy use is a reasonable indicator of the scale and intensity of human activities on our planet.

The vehicle set off at a speed of one kilometre per hour 12 hours ago.

Four hours ago, it picked up speed and was travelling at 30 km/hr.

One hour ago it was going at 100 km/hr.

Fifteen minutes ago at 350 km/hr.

Six minutes ago at 1,000 km/hr.

Three minutes ago at 3,000 km/hr.

It is now traveling at around 20,000 km/hr.

Visibility is not good — and we, the passengers, don't have a clear view of where we are going. But among us there are some scientists who have made a study of the environment, and they are warning that we are heading for a precipice. They are shouting out to us to slam on the brakes and change direction.

But most of us, especially those in charge, are hell-bent on making our vehicle go faster than ever.

Source: Stephen Boyden

The main sources of extrasomatic energy throughout the industrial phase of society have been fossil fuels, although the relative contributions of coal, oil and natural gas have changed over the past 60 years. In some countries, nuclear power now makes a significant contribution to the generation of electricity.

The last 1,000 years

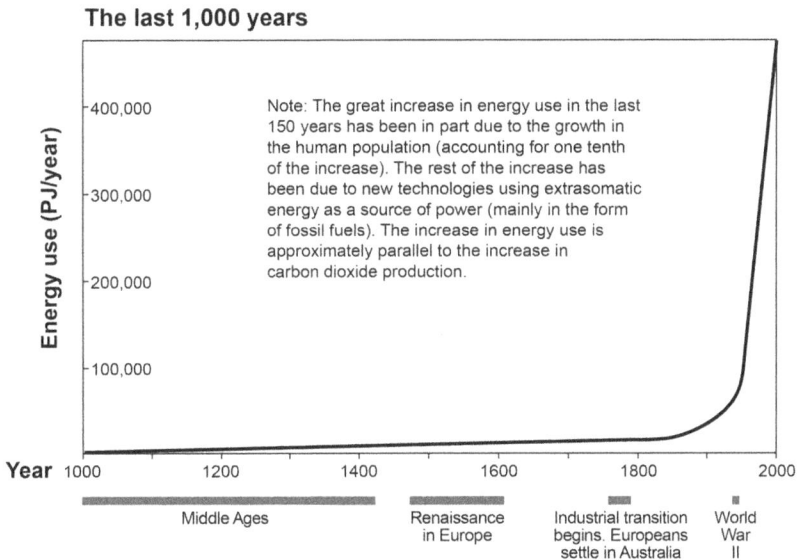

Note: The great increase in energy use in the last 150 years has been in part due to the growth in the human population (accounting for one tenth of the increase). The rest of the increase has been due to new technologies using extrasomatic energy as a source of power (mainly in the form of fossil fuels). The increase in energy use is approximately parallel to the increase in carbon dioxide production.

Figure 5.2 Energy use by the human species

Source: Stephen Boyden

Hydroelectricity, unlike fossil fuels and nuclear power, does not produce undesirable by-products, and it makes a significant contribution to the available power in regions where the topography allows it. Use of other clean, non-polluting energy sources, such as wind and solar power, are on the increase but, so far, they contribute only a small fraction of the total energy budget.

Other technometabolic inputs into human societies today include a vast range of materials used in the construction of buildings and roads and for the manufacture of machines and utensils as well as electronic devices. To take just one example, the per capita consumption of iron in Australia today, excluding the iron in manufactured goods imported from overseas, is around 1.3 kilograms per day. In Shakespeare's time it was probably about one gram per day.

Technometabolic outputs: Carbon dioxide and climate change

If it were not for certain gases occurring naturally in the atmosphere, the world's average temperature would be 33°C colder than it is. That is, it would be around −18°C instead of 15°C.

This is because these gases trap some of the infrared radiation that escapes from the Earth's surface. This blanketing effect results in the lower layers of the atmosphere being warmer, and the upper layers colder, than would be the case if these gases were not there. This phenomenon is known as the natural greenhouse effect.

Water vapour is responsible for about 80 per cent of the natural greenhouse effect. The remainder is due to carbon dioxide, methane, and a few other minor gases.

Carbon dioxide (CO_2) is responsible for about 15 per cent of the natural greenhouse effect. This is to say that were it not for the CO_2 in the atmosphere, the Earth's average temperature would be 5°C cooler than it is.

For the first 200,000 years of the history of modern humans (*Homo sapiens*), the mixture of these natural greenhouse gases was relatively constant. During the past 200 years, however, there has been an increase in the CO_2 concentration in the atmosphere resulting from human activities — from 292 parts per million to 400 parts per million in 2016. This increase in atmospheric CO_2 is mainly the result of two sets of human activities: (1) deforestation, and (2) the combustion of fossil fuels as a source of energy for driving machines and providing heat.

The amount of carbon dioxide emitted by the human population today is around 10,000 times greater than it was when farming began some 450 generations ago, and over 90 per cent of this increase has occurred over the past 100 years (Figure 5.3). It is predicted that the concentration of carbon dioxide in the atmosphere will reach double the pre-industrial level by 2050.

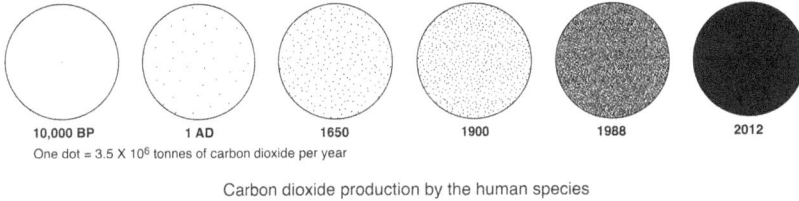

| 10,000 BP | 1 AD | 1650 | 1900 | 1988 | 2012 |

One dot = 3.5 X 10^6 tonnes of carbon dioxide per year

Carbon dioxide production by the human species

Figure 5.3 Carbon dioxide production by the human species
Source: Stephen Boyden

As a consequence of this increase in atmospheric CO_2, the Earth's average surface temperature has increased by about 0.8°C, with about two-thirds of the increase in temperature occurring since 1980. This is known as the enhanced greenhouse effect.

If all populations around the world had the same intensity of technometabolism as the developed countries, the increase in CO_2 emissions since the time that farming began would be around 50,000 fold.

From these facts, it would seem very likely that the continuing anthropogenic increase in the concentration of CO_2 in the atmosphere will result in a significant increase in global temperature. If no action is taken, the consequences for humanity will be very serious.

Because of the complexity of the carbon cycle, however, there are uncertainties about the precise effect of increasing CO_2 in the atmosphere on global temperature. For example, it cannot be assumed that doubling the CO_2 concentration in the atmosphere will simply double the contribution of this gas to global warming; that is from around 5°C to 10°C — an extra 5°C.

Mathematical models have been designed to predict the likely increase in temperature due to increasing concentrations of carbon dioxide. Because of the uncertainties in the system, these models come up with different results that range from a further increase in the 21st century of 1.1°C to an increase of 6.4°C.

There are also differences of opinion within the scientific community about the relative contributions of fossil fuel use and deforestation to the increase in CO_2 in the atmosphere over the past 250 years. One view holds that most of the rise in atmospheric CO_2 since 1750 has been due to the destruction of the capacity of forests and soils to take up CO_2 from the atmosphere rather than the use of fossil fuels.

As in all reform movements, there is a predictable backlash from counter-reformers. In the case of climate change, the counter-reformers are commonly referred to as climate change deniers. The main disputed issues relate to the causes of the increase in average global temperature, whether humankind is responsible for it, and what will be the likely consequences of global warming.

Here are a couple of typical quotations from two of the most vociferous climate change deniers of our time:

> I am convinced that policies meant to reduce carbon dioxide-induced global warming will be destructive The right response to the non-problem is to have the courage to do nothing ... Climate change is a non-problem. Even if the higher estimates of global sensitivity were correct, there is no hurry to take any action.
>
> — Lord Monkton[6]

> Man-made climate change has become one of the most dangerous arguments aimed at distorting human efforts and public policies in the whole world ...
>
> Climate change is caused not by human behaviour but by various exogenous and endogenous natural processes (such as fluctuating solar activity).
>
> — Václav Klaus, former president of the Czech Republic

Klaus describes concern about climate change as a 'new wave of dangerous indoctrination of the whole world' and says that 'global-warming alarmism is challenging our freedom'.[7]

However, 97–98 per cent of the most published climate researchers believe humans are causing global warming, and the finding that the average global temperature has increased in recent decades as a result of human activities has been endorsed by the academies of science in all the major industrialised countries.

---

6   www/ossfoundation.us/projects/global warming/myths/christopher-monckton.
7   www.guardian.co.uk/environment/georgemonbiot/2009/mar/o6/climate-change-deniers-top-10.

Other greenhouse gases

Other greenhouse gases are on the increase in the atmosphere as the result of human activities. The concentration of methane has increased 2.5 times since the beginning of the industrial era. Although it is in much smaller quantities than carbon dioxide, it is 21 times as effective per molecule and this gas now contributes 20 per cent of the enhanced greenhouse effect. It is estimated that 64 per cent of the methane emitted into the atmosphere today is the result of human activities. Of these human-induced emissions, 33 per cent are the result of the use of fossil fuels, 27 per cent come from fermentation in gastrointestinal tracts of cattle and some other farm animals, and 16 per cent from the decomposition of organic matter in land fill.

Other gases contributing to the enhanced greenhouse effect are CFCs (chlorofluorocarbons, see below), ozone and nitrous oxide.

A recent development

In December 2015, the United Nations held a conference on climate change in Paris. The conference, which was attended by representatives of 195 countries, agreed to set a goal of limiting global warming to less than 2°C higher than pre-industrial levels. The agreement calls for zero net anthropocentric greenhouse gas emissions to be reached during the second half of the 21st century. Furthermore, the parties agreed to 'pursue efforts to' limit the temperature increase to 1.5°C.

This conference is indeed a most important step forward, although it remains to be seen to what extent the countries across the world comply with its recommendations.

CFCs

There has been a steady accumulation over recent years of chlorofluorocarbons (CFCs), methyl bromide and halons in the atmosphere. CFCs are synthesised chemical compounds used in refrigerators, and methyl bromide is a biocide used for the control of insect pests in the soil and in grain products. The main use of halons is in fire extinguishers.

In 1974, Frank Rowland and Mario Molina suggested that these compounds might, on reaching the stratosphere, destroy the ozone that protects the surface of the Earth from ultraviolet (UV) radiation from the Sun. This would result in increasing damage to terrestrial organisms, with serious consequences for the natural environment and for humankind. The shorter wavelength UV-B rays are especially harmful. It is predicted that the yields of soybeans, peas and beans will decrease by a quarter if UV-B radiation increases by 25 per cent. Increase in UV radiation is also likely to destroy plankton at the surface layers of the oceans. Because phytoplankton are at the base of the oceanic food chain, this change will have a devastating impact on populations of fish and other animal life in the sea.

Eventually, evidence was forthcoming that the ozone layer was indeed thinning. This evidence was promptly disputed by representatives of the aerosol and halocarbon industries. For instance, the chair of the board of DuPont was quoted as saying that ozone depletion theory is 'a science fiction tale … a load of rubbish … utter nonsense'.[8]

In this case, however, the counter-reform backlash was relatively short-lived. The processes of cultural reform are now well advanced and, as a result of international agreements, there has been a major reduction in the release of CFCs and related compounds into the atmosphere, and it is now hoped that the ozone layer will be back to normal by around 2065.

Persistent organic pollutants (POPs)

In *Silent Spring*, Rachel Carson drew attention to the insidious and destructive ecological impacts of DDT, which is one of a group halogenated hydrocarbons that are used as pesticides and in various technological processes and which have become known as persistent organic pollutants (POPs). POPs accumulate in the internal organs of living creatures and are believed to be responsible for increasing and widespread infertility in wild animals, and probably also humans. They are also suspected of contributing to the increase in breast cancer in women and to reduced sperm counts in men. POPs are persistent in the natural environment and have been found in the organs of animals in areas that are far away from where they were originally released, such as the Arctic and the Antarctic.

---

8   www.en.wikipedia.org/wiki/Montreal.Protocol#History.

Some authorities consider pollution of the environment with POPs as being as serious a problem for life on Earth as the enhanced greenhouse effect or the thinning of the ozone layer.

A strong counter-reform backlash occurred in response to Carson's claims. To take just one example, the president of a large chemical corporation described her as 'a fanatic defender of the cult of the balance of nature'.[9]

Local air pollution

Local air pollution with hydrocarbons is an important ecological issue in many urban areas, especially in Asia — Beijing is a notorious example. The main cause is the combustion of fossil fuels in power stations, factories and motor vehicles.

Particles of less than 10 micrometres (PM10s) and those less than 2.5 micrometre (PM2.5s) are especially important, being small enough to penetrate deeply into the lungs. These pollutants can cause respiratory disease in humans, including pneumonia, bronchitis and asthma.

Technoaddiction

In the history of civilisation it has frequently been the case that new techniques have been introduced simply for curiosity, or sometimes because they have benefited a particular individual or group within society. But, with the passing of time, societies have organised themselves around the new techniques and their populations have become progressively more and more dependent on new technologies for the satisfaction of basic needs. Eventually, a state of complete dependence is reached.

The dependence of the populations of ecological Phase 4 societies on fossil fuels is an obvious and serious example. Others include our dependence on electricity and, quite recently, on computer technology.

---

9   P.R. Erhlich, A.E. Erhlich & P.R. Holdren, 1977. *Ecoscience: Population, resources, environment*. 2nd edn. W. H. Freeman, San Francisco, CA, pp. 854–56.

This insidious form of addiction passes largely unnoticed, although it is often of immense economic and ecological significance.

From the ecological standpoint, it is significant that in the modern cultural setting the following basic behaviours generally use up much more energy and create much more pollution than they did in the past: seeking in-group approval, seeking to conform, seeking attention, seeking novelty, seeking excitement, seeking variety, seeking comfort, visiting relatives, being selfish, being greedy and being generous.

Ecological Phase 4 will soon come to an end

We don't have to be ecologists to appreciate that the living systems of our planet that support us will not be able to tolerate this relentless maltreatment from the human species ad infinitum. At present, anthropogenic climate change is the most urgent threat. But there are many other critical issues that require urgent attention if civilisation is to survive (Box 5.2). If present trends continue unabated, the collapse of civilisation is inevitable. The days of ecological Phase 4 are numbered.

The most disconcerting feature of the present situation is the fact that the prevailing cultures of the world are blissfully unaware of these ecological realities. They incorporate powerful delusions that are incompatible with the achievement of ecological sustainability and therefore the survival of civilisation. They have lost sight of our total dependence on the life processes that underpin our existence, and they have no grasp of the magnitude and seriousness of current human impacts on the ecosystems of our planet.

Box 5.2 Some serious signs of cultural maladaption in the modern world

- A steady and continuing increase in the concentration in the atmosphere of the greenhouse gas carbon dioxide, from the pre-industrial level of 280 parts per million by volume to 400 parts per million in 2013. This is due to the use of fossil fuels as a source of energy by humankind and to widespread deforestation. There is strong evidence that this change is leading to increased temperatures across the globe and to other climatic disturbances. If allowed to continue unabated it could lead to a massive drop in the global population later in this century.

- Destruction of 80 per cent of the world's original forests. At present, trees are felled in the Amazonian forests at the rate of 2,000 a minute. Deforestation is contributing to climate change and is resulting in great loss of biodiversity.

- Severe land degradation (due to loss of organic matter, disruption of natural nutrient cycles, soil erosion and salinisation) resulting from deforestation and unsatisfactory farming practices. According to the FAO, a quarter of farming land is highly degraded. Another 8 per cent is moderately degraded and 36 per cent is classed as stable or slightly degraded. Ten per cent is described as 'improving'.

- Worldwide loss of biodiversity on land and in the oceans. According to some estimates, 25 per cent of all mammal species could be extinct in 20 years' time.

- Persistent organic pollutants (POPs) are now found in the tissues of humans and other animals all over the world. POPs are synthetic compounds used as pesticides and for other purposes. They can cause ill health or death and they interfere with reproductive processes.

- Acidification of the oceans resulting from an increased uptake of carbon dioxide from the atmosphere.

- Thousands of weapons of mass destruction stored in the arsenals of the world — many times more than necessary to bring an end to the human species.

- Violent conflicts across the world between people holding different beliefs about the supernatural.

- Extreme disparities in health and material wealth among human populations (this was not the case for the first 190,000 years of human existence).

Source: Stephen Boyden

6

Religion and warfare in biohistorical perspective

Both religion and warfare loom large in the recorded history of humankind, and they share some important features. They have both had immeasurable impacts on the life experience of vast numbers of humans, and they both illustrate the amazing power of culture to shape people's mindsets and determine their behaviour, sometimes in ways that are very much to their disadvantage. They are also linked by the fact that warfare is sometimes based on disagreements between different religious groups about the nature of a deity or the authenticity of various prophets.

Both warfare and religion have caused an immense amount of human suffering. Religion, however, unlike warfare, is also a source of great comfort for some people.

Religion

An early outcome of the capacity for culture was the emergence of religion as a universal feature of human society. Without exception, all recent hunter–gatherer groups have embraced belief in a supernatural, or spiritual, dimension of the universe. While there was enormous variation in the details of these belief systems, they all involved belief in spirits or gods, and they all provided a religious explanation of human existence. There is every reason to suppose that this was the case for many tens of thousands of years.

Early farming and early urban societies were characterised by powerful religious beliefs. All early farming civilisations worshipped gods and spirits. Although the details differed from one region to another, the dominant religious theme for several thousand years in south-western Asia and Europe was the notion of a mother goddess or 'female principle', who was worshipped as the giver of life. Rituals were aimed at pleasing the goddess in the hope of improving the chances of good harvests, good health and successful reproduction.

In the early cities of Mesopotamia, the religious sense of oneness with nature was abandoned for a sense of separation. Each city state had its own god, who was now male, and conflicts between city states were viewed as being conflicts between the different gods.

The complicated story of the developments in, and interactions between, different religions in the Early Urban Phase of human history is well beyond the scope of this chapter. A few points are, however, worth making from the biohistorical perspective.

Most of the religions of the Near East were polytheistic, and this was also true in ancient Greece and Rome. However, Zoroastrianism, which was founded by the Persian prophet Zoroaster in the late 7th or early 6th century BC, was based on the idea of a continuous struggle between a single god of creation, goodness and light, and his archenemy, the spirit of evil and darkness. Unlike some other early urban religions, Zoroastrianism included a highly developed ethical code. Judaism was also based on belief in only one god.

The teachings of Buddha around 600 BC were initially relatively simple, as were those of Jesus of Nazareth. In neither case, however, was this simplicity to last. In the case of Christianity, the process of elaboration, intellectualisation and institutionalisation soon led to complicated sets of theories and rituals, with notable contributions from older religions and philosophies. A professional priesthood came into being and, ultimately, the Christian Church split into a few large and often mutually intolerant sects, and numerous smaller ones.

Another great religion of early civilisation, Hinduism, became extremely complex and involved the worship of a few major deities and countless minor ones, although the various sects within Hinduism were relatively tolerant of each other and of other religions. Similar elaboration occurred in the religions of the New World before the European conquest. In the case of the Toltecs, who dominated the valley of Mexico before it was overrun by the Aztecs, a relatively basic nature-worship was transformed into an elaborate polytheism.

Despite all the diversification and other changes that took place in most religions, the great majority of humans were born, grew up and lived their lives in cultural systems that clearly defined the nature of a supernatural world, and which spelled out their religious obligations. Most of these people never doubted the validity of their particular belief system and they assumed, therefore, that all other religions were wrong. There were exceptions, of course, as in the case of those individuals who became suddenly converted from one religion to another after some kind of revelation or an encounter with a charismatic teacher. And there have always, no doubt, been a few sceptics.

A strong sense of religious conviction and differences in religious beliefs have, throughout the history of civilisation, been the cause of a vast amount of bloodshed and untold suffering among many millions of humankind. In the case, for example, of Islam and Christianity, mutual intolerance led to a great deal of bloodshed, as in the case of the Crusades. Indeed, Mohammed achieved his initial success in establishing Islam as the dominant religion over the whole of Arabia through a series of military engagements with local populations of different religious persuasions. But religious intolerance was not confined to the Western world. In China, for instance, religious persecutions around 845 AD are said to have resulted in the destruction of 44,600 Buddhist religious establishments and the enslavement of 150,000 Buddhist nuns and monks.

The assumption in military conflict that one's own god is on one's side has persisted for several thousand years. A Spanish eyewitness to the conquest of Middle America wrote in his diary: 'When the Christians were exhausted from war, God saw fit to send the Indians smallpox, and there was a great pestilence in the city.'[1]

Occasionally, of course, there have been individuals who have been able to perceive the insanity of religious bigotry. Akbar, a Mogul emperor of India from 1556 to 1605, was such a person. He became acutely aware of the absurdity of the multifaceted and fragmented religious scene in India, and of the needless distress and wasteful destruction caused by religious intolerance. He refused to accept the idea that because he, the conqueror and ruler, happened to have been born a Mohammadan, therefore Mohammadism was true for all humankind. It was his aim that all people, whatever their race or religion, should participate equally in India's public life.

Today, the vast majority of the world's population adheres to one religion or another. One estimate suggests that about 33 per cent of people are Christians; 19.6 per cent, Muslims; 13.4 per cent, Hindus; and 5.5 per cent, Buddhists. There are also countless different sects within the major religions, each with a particular creed.

Judaism, Sikhism, Zoroastrianism and Taoism are among numerous other smaller religious groupings, each of which is adhered to by less than 1 per cent of the total human population. Only about 15 per cent of people are described as non-religious.

As in the past, some religious groups are so intolerant of those that hold somewhat different beliefs about the supernatural that they set about slaughtering them. It would be difficult to find a single edition of a major daily newspaper today that does not report at least one tragic consequence of this horrendous and senseless behaviour.

---

1   D.R. Hopkins, 1983. *Princes and peasants: Smallpox in history.* University of Chicago Press, p. 201.

Warfare

Killing members of one's own species is not common among mammals. It occurs occasionally among chimpanzees and some other species, including wolves and the big cats. Large-scale and deliberate killing of one's own kind is, however, a uniquely human characteristic. Perhaps the nearest thing among other animals is the aggressive behaviour that is sometimes seen between colonies of certain species of ants.

Judging from evidence of recent hunter–gatherers, mortal conflict sometimes occurred between different hunter–gatherer bands — although it was not a constant feature of primeval society. Many groups lived at peace with their neighbours for long periods.

Large-scale and highly organised homicide has been one of the hallmarks of civilisation. Some authors have suggested that it began with the domestic transition, when people came to possess commodities, like animals and stored grain, which were coveted by others. While there may well be something in this idea, organised violence between groups is certainly not a necessary outcome of agriculture. There have been plenty of farming communities that have lived at peace with their neighbours for long periods of time, and archaeological evidence suggests that the early farmers of the valleys of the Tigris and Euphrates rivers, as well as those of central Europe, were not involved in warfare.

On the other hand, it is also clear that, in some farming societies, violent hostilities were an important aspect of life from very early times. Towards the end of the Neolithic Phase in Europe, the relative peace was shattered by the aggressive 'battle-axe' people, who were intent on warfare and political domination; and there is evidence that, around 5,000 years ago, farming people in the south of England built fortified settlements. In more recent times, several groups of slash-and-burn agriculturalists in South America, such as the Yanomami, have been almost constantly at war with their neighbours.

It has also been suggested that the fact that villagers in some regions were usually at war with their neighbours, while in others they were not, can be explained by differences in local ecological conditions. Competition for scarce resources is seen to be the underlying cause of conflict.

My view is that the chief determinants lay in differences in cultural, rather than biophysical, systems, although both factors probably played a role.

It was soon after the establishment of the early cities in Mesopotamia that organised violence between large groups of people came to be accepted as normal. According to one view, these cities came into existence in response to increasingly frequent attacks on villages by nomadic peoples from outside the valley, and the people collected together under temporary military leaders for self-defence. Another school of thought argues that the converse is true, and warfare was the result of the existence of cities, and that the material wealth accumulated within them encouraged attacks from plundering bands of barbarians.

Whatever the explanation, by around 5,000 years ago, highly organised fighting among the city states, and between city states and barbarian raiders, was commonplace. People were immersed in cultural systems that glorified the military exploits of their forefathers and that characterised other human populations as enemies. The heroes of society were the successful commanders and intrepid warriors. For centuries, history books have extolled the prowess of men who commanded armies or navies that succeeded in annihilating large numbers of perceived enemies. Plutarch, for example, a writer who is otherwise generally known for his humanitarianism, wrote in the following glowing terms of Julius Caesar:

> Caesar surpassed all other commanders in the fact that he fought more battles than any of them and killed greater numbers of the enemy. For, though his campaigns in Gaul did not last for as much as ten complete years, in this time he took by storm more than 800 cities, subdued 300 nations and fought pitched battles at various times with three million men of whom he destroyed one million in the actual fighting, and took another million prisoners.[2]

Nevertheless, from the beginning, warfare usually had to be 'justified'; wars were waged in the name of a god or, at least, of an empire, which was depicted as bringing peace or other benefits to the conquered peoples. The following words from the Book of Joshua provide a good example of the former:

---

2   E. Canetti, 1973. *Crowds and power.* C. Stewart (trans.) Penguin, Harmondsworth, p. 269.

And Joshua at that time turned back and took Hazor, and smote the King thereof with the sword: for Hazor beforetime was the head of all those kingdoms. And they smote all the souls that were therein with the edge of the sword, utterly destroying them: there was not any left to breathe: and he burnt Hazor with fire. And all the cities of those kings, and all the kings of them, did Joshua take and smote them with the edge of the sword, and he utterly destroyed them, as Moses the servant of the Lord commanded.[3]

The complete lack of compassion for the people who made up the enemy is well illustrated by descriptions from the tombs of some of the Pharaohs of the New Kingdom of Egypt. The army of Merneptah, son of Rameses II, won a great battle against the Libyans. The booty included not only 9,376 prisoners, but also the genitalia of all the dead enemy soldiers (or their hands, if they had been circumcised). These items were loaded onto donkeys and brought home by the victorious soldiers as evidence of their success. Rameses III also waged war against the Libyans and, in this case, the booty contained 12,535 of these human parts. Clearly, in the prevailing cultural setting, people found this behaviour to be entirely acceptable and praiseworthy.

The professional soldier came to be accepted as a natural and necessary component of most urban societies. For millennia the sword held pride of place among human artefacts as the symbol of masculine virtue.

Warfare was not, however, an inevitable concomitant of urbanisation. The remains of the township of Caral in Peru, which came into existence around 5,000 years ago, show no trace of warfare. No battlements and no weapons have been found. Similarly, excavations at ancient cities of Harappa and Mohenjo-daro in the valley of the Indus River in Pakistan have revealed no indications of military activity until the very end of their history. Another striking example is provided by the cities of the Minoan civilisation on the island of Crete. At a time in the Bronze Age when warfare and empire-building were gaining momentum throughout the Near East, the

---

3   Joshua 2:10–12 (King James Version).

people of Minos were creating 'one of the most graceful civilisations man has ever achieved'.[4] The enthusiasms of the people of the Minoan city of Knossos were athletics, elegant clothing and the natural world:

> Most remarkable of all is the apparently quite peaceful nature of this development over many centuries, testifying to a highly stable social system, confident, dynamic and flexible. Lack of interest in pictures of battles and of warriors generally is very remarkable. Weapons are rarely found in Cretan tombs before the Late Minoan II level, but they are common in Mycenaean mainland.[5]

This state of affairs presumably had something to do with the fact that Crete is an island. If the Minoan cities had been on the mainland, military activities would surely have been necessary, if only as a means of self-protection.

There have been occasional individuals who have personally renounced violence in all its forms. An especially interesting example was Ashoka, who discarded a strong cultural tradition of violence to embrace the cause of peace. He inherited an empire that extended from Afghanistan to Mysore in India, and which had been built by his grandfather through military force. Around 262 BC, Ashoka was in the process of further extending this empire when he found himself involved in a war with the Kalinga people. His army was victorious, and 100,000 persons were slain. This experience of war, however, brought about a remarkable change in Ashoka. Suddenly he became acutely aware of the intensity of human distress caused by the fighting, and he adopted non-violence as the creed of his life. He changed his personal religion and converted to Buddhism, which, of all the religions of India at that time, was most strongly identified with the principle of Ahisma.[6]

Ashoka issued a long series of religious edicts, which were written on rocks and pillars. At one point he was able to say, 'instead of the reverberation of the war-drum, is now to be heard the reverberation of religious proclamations'. According to one of the edicts, he was

4   L. Woolley, 1963. 'The beginnings of civilisation: Part 2', in J. Hawkes & L. Woolley (eds), *Prehistory and the beginnings of civilisation, vol. 1: History of mankind: Cultural and scientific development*. George Allen and Unwin, London, p. 241.
5   R.F. Willetts, 1965. *Ancient Crete: A social history from early times until Roman occupation*. Routledge Kegan and Paul, London, pp. 128–29.
6   Ahisma is the law of reverence for, and non-violence toward, every form of life.

anxious to ensure that 'his sons and grandsons may not think it their duty to make any new conquests'. Later in his reign, he extended the principle of non-violence to animals, and a decree was issued prohibiting the slaughter of numerous specified birds and beasts. He also established botanical gardens especially for the cultivation of plants, herbs, roots and fruits for medicinal purposes, and he arranged for the establishment and maintenance of hospitals both for humans and for animals. In one of the edicts, he wrote:

> On the roads, too, banyan trees have been planted by me to give shade to man and beast; mango-gardens have been planted and wells dug at every half-kos; rest houses, too, have been erected; and numerous watering-places were made here and there for the comfort of man and beast.[7]

Non-violence did not become important in the religions of the Near East before the teachings of Jesus Christ, but after that time it was probably believed in and practised by Christians for a few hundred years. By the 4th century AD, however, cultural developments had overridden the fundamental tenets of the teaching of Jesus, and large-scale homicide was once again acceptable, so long as one was fighting on the side of God.

One of the profoundly significant consequences of the human capacity for culture has been the transmission of hatred across generations. This may well have happened sometimes in hunter–gatherer societies, but it became much more important as an influence on human affairs after the development of civilisation, and it is a major determinant of human behaviour in many parts of the world today. A typical sequence of events is as follows: two human groups, who recognise each other as being different in terms, for example, of religious beliefs, language or skin colour, come in contact. As a result of some aggressive act on the part of one of the groups, or perhaps a misunderstanding, the natural mutual suspicion between the two groups escalates to overt hostility, and eventually violence. This, in turn, leads to distress, resentment and anger among the survivors of both groups. This common pattern is serious enough in its own right, but it is of minor significance

---

7   These quotations about Ashoka are from R. Mookerji, 1928. *Asoka*. Gackwad lectures. Macmillan, London, pp. 21–22.

compared with the monstrous tragedy that descendants of the two groups, many generations later, are forced to feel the mutual hatred and to continue the violence. Such is the power of culture.

Weaponry

Early in hominid history, our ancestors applied their toolmaking prowess to the manufacture of weapons. In primeval times, these were used mainly in hunting animals for food, although they may have been used on occasion in combat between human groups. The weapons were broadly of two classes. First, there were close-range weapons, like clubs and hand axes, which consisted in essence of an extension of the human arm or hand. They were used for directly striking the enemy. Second, there were projectile weapons, such as stones, sticks and boomerangs, which were thrown at the target, initially by the human arm, but later by other means, as in the case of the bow and arrow. Spears were used as both close-range and projectile weapons.

After the beginning of urban civilisation, most weapons were designed especially for killing people, and they fell into the same two classes: close-range and projectile weapons. There was a simultaneous development of armour, made of leather or metal, which was intended to provide soldiers with some protection against the weapons of the enemy.

The spear is the most ancient of the weapons used in warfare. In one form or another it had been used for tens of thousands of years in hunting animals for food, originally with a shaft of wood and a spearhead of stone. When techniques of metallurgy were developed, spearheads, and sometimes spear shafts, were made of copper or bronze. The soldiers of Sumer, 5,000 years ago, and of the Old Kingdom of Egypt, 4,700 years ago, were equipped with metal spears. The cavalry version of the piercing spear, the lance, was developed later.

The other important short-range weapon, invented and developed especially for cutting or thrusting into flesh, was the sword. This consisted of a pointed blade, which might be straight or curved, with a handle, or 'hilt', and a cross-guard. One or both edges of the blade were usually sharp. Soldiers often became emotionally attached to their swords and even gave them names. Famous examples from history and legend are Charlemagne's 'Joyeuse', and King Arthur's 'Excalibur'.

The discovery of the explosive potential of a mixture of saltpetre, sulphur and charcoal, otherwise known as gunpowder, is believed to have been made in China over 1,000 years ago. In the mid-13th century, Roger Bacon in England wrote a formula for gunpowder as follows: seven parts of saltpetre, five parts of young hazel wood (charcoal) and five parts of sulphur. He stated that this mixture would explode, and that it could cause an enemy to be blown up, or at least to flee in terror. It is believed that the Moors used gunpowder in warfare around 1250, by putting a kilogram of the explosive mixture into an iron bucket that had a small touch-hole at the bottom. They placed a pile of stones on top of the gunpowder, which was then ignited. The resulting explosion propelled the stones through the air, ideally towards the target.

The first cannons, which were made of bronze, were introduced at the beginning of the 14th century. They were replaced by iron cannons half a century later. The first military event of importance in Europe in which artillery played a significant part was the capture of Constantinople by the Ottoman Turks in 1453.

Gunpowder was also applied to the development of guns to be held in the hand, but for a long time these were ineffective. This was partly because of the need to keep a match alight in the combat situation, partly because of the difficulty in keeping gunpowder dry, and partly because of the clumsiness of stuffing lead bullets into the gun's barrel with a ramrod. Effective rifles were not used for military purposes until the Thirty Years War (1618–48). Later technological developments greatly increased the accuracy and range of both cannons and handguns.

Cultural evolution in Europe and Asia over the past 5,000 years has been associated with a progressive increase in the number of people actively participating in, or affected by, wars. In the 1914–18 Great War, about 53 million men were mobilised into the armed forces, and 8 to 10 million were killed. In 1914, 640,000 French soldiers lost their lives during the four months from August to November. The army of the United Kingdom lost over 400,000 men in the battle of the Somme, 50,000 of them on the first day.

The Second World War differed from previous conflicts in that aerial bombing of important cities resulted in large numbers of civilian casualties. The armed forces of the warring nations numbered about 30 million and the total number of individuals killed, military and civilian, was probably between 35 and 40 million.

To the time of writing, a further world war has so far been averted, although serious military conflicts have taken place in Europe and Asia, resulting in hundreds of thousands of deaths.

Not all deaths in warfare have been due to physical injuries inflicted by enemies. Malnutrition and infectious disease have also taken their toll. The story of the Spanish invasion of Mexico under Hernando Cortez early in the 16th century provides an interesting example. One of the Spanish expeditions that landed at present-day Vera Cruz in April 1520 included an African slave who was infected with smallpox. The disease soon appeared in the native population and, because these people had not had any previous contact with the smallpox virus, it spread extremely rapidly. By September of that year, the disease had reached the towns around the lakes in the Valley of Mexico, including the Aztec capital, Tenochtitlan. About half the population of this city and of the surrounding region died within six months. This happened at a time when the Aztecs had been gaining the upper hand in the conflict with the Spanish forces. Because most of the Spaniards were immune to the disease, however, they were able to exploit the situation to their military advantage, eventually overcoming the indigenous armies.

This was not by any means the only military campaign in which infectious disease played a role. Until very recently, microbes have caused more deaths among warriors than combat itself, as reflected in the following figures:

- *Crimean War* (1853–56) — about 60,000 men on both sides killed or died of wounds, about 130,000 died of disease.
- *American Civil War* (1861–65) — about 220,000 men killed or died of wounds, about 400,000 from infectious disease.
- *South African War* (1899–1902) — of the British forces, 7,534 were killed or died of wounds, 14,382 died of infectious disease.

The influenza epidemic immediately after the First World War killed at least 21 million people, and probably many more, compared with the 8 to 10 million soldiers killed in action.

The increasing power of weaponry has led to a fundamental change in the 'art of warfare'. By the time of the First World War, some of the combat was no longer on a one-to-one basis. The single touch of the trigger of a machine gun could kill a dozen men, and one artillery shell could destroy many individuals who were out of sight of the gunners. This change in the nature of armed conflict had further progressed by the time of the Second World War, when there was greatly increased air bombardment and new projectile weapons in the form of various kinds of rockets were introduced. Technology now exists that makes it possible for deliberate hostile action by a small number of individuals to cause the death of millions of people thousands of kilometres away.

During the last part of the Second World War, Germany and the United States were competing to be the first to produce nuclear weapons. Then, three months after the capitulation of Germany, on 6 August 1945 at 8.15 am, a nuclear bomb was dropped from an American aircraft onto the city of Hiroshima in Japan. At least 140,000 people, about 40 per cent of the population of the city, were immediately killed or died soon afterwards. The buildings of the city were flattened over an area of 13 square kilometres. Three days later, another bomb was dropped on the Japanese city of Nagasaki, and 26 per cent of its population of about 280,000 was killed outright.

After that time, governments representing opposing political ideologies directed immense financial resources and human effort to the development of nuclear weapons, with the result that bombs now exist with an explosive power a thousand times greater than that which was dropped on Hiroshima. Nuclear weapons range in strength from the equivalent of around 100 tonnes to 20 million tonnes of TNT, depending on the use for which they are designed.

The ecological and human impacts of a nuclear conflict in the future would obviously depend on the scale of the war and on the geographical distribution of the nuclear explosions. Certainly, even if only one tenth of the existing nuclear weapons were used, the numbers of people killed by radiation, fire and blast would be astronomical. While most commentators consider it likely that a major nuclear war would leave some survivors, especially in the Southern Hemisphere, uncertainties exist about the likely effects of such a war on the planet's ecosystems resulting from nuclear radiation and climate change caused by widespread fires. It could well happen that the biosphere as we know it today would collapse and no longer be capable of supporting a human population.

It has thus come about that, for the first time in the history of life on Earth, and in the lifetime of many of us alive today, a single species of animal has developed the ability to destroy most, if not all, of its kind within a few days, and to cause extreme devastation in the biosphere as a whole. It owes this achievement to its capacity for culture.

Mention must also be made of the enormous amount of effort and resources that have been devoted in modern Exponential Phase societies to the development of other sophisticated weapons of mass destruction. Thus, apart from the advances in nuclear armaments, great progress has been made in the development and production of chemical and biological weapons. I will not discuss these here because, horrendous though they may be, their impact on humans and other life forms would be small in comparison with that of a nuclear war.

Human society as a whole spends over US$1 million per minute on the development and manufacture of homicidal devices. In six hours, more money is spent on the manufacture of arms than was spent by the world community in bringing about the eradication of smallpox from the face of the Earth.

The facts summarised in the last couple of pages make sheer mockery of the scientific name that humans have given themselves — *Homo sapiens*. Yes, humans have big brains, they are clever, and they have a capacity for culture. But they can hardly be described as wise.

The story of warfare, of the development of nuclear weapons, and now of the rise of international terrorism, well illustrates the potential of culture to lead us to behave in ways that are nonsensical in the extreme. Of course, most of the individual humans who have, for example, participated in the manufacture of nuclear weaponry, have been behaving in a moderately rational way, in terms of the assumptions of their particular cultural microcosms. But their behaviour can in no way be seen as rational in terms of the well-being of humanity or of the living systems of the biosphere.

In conclusion, there is nothing in human nature that precludes the performance of extremely aggressive and brutal acts directed by groups of people against other groups of people. On the other hand, there is also nothing in human nature that rules out the possibility of different human groups living at peace with each other. With regard to the future, a major determinant of whether or not warfare and terrorism continue to be a feature of civilisation will be the extent to which people allow themselves to be blinded by narrow, pernicious and maladaptive cultural delusions.

Box 6.1 Increasing destructive power of weaponry

The growth in the killing potential of bombs during the 20th century can be illustrated by the following analogy: If we imagine the explosive power of the biggest bombs in the First World War to be represented by a pea, then the most powerful weapons used in the Second World War (other than atomic bombs) would equal the size of a large plum. The Hiroshima bomb would be equivalent to a sphere of about 0.5 metres across. The most powerful bombs that are now ready for use would have a diameter of five metres.

Source: Stephen Boyden

7

Healthy people on
a healthy planet

The best hope for the future lies in a fifth watershed in cultural evolution, leading to a new ecological phase in human history — a phase that is based on understanding the human place in nature and in which human society is sensitive to, in tune with, and respectful of the processes of life. This is termed a biosensitive society.

The word biosensitive fulfils the need for a single word to describe a society with these characteristics. The expression 'ecologically sustainable' has come to be used widely in recent years. Of course, society must be ecologically sustainable — otherwise, in the long term, it cannot continue to exist. But ecological sustainability is surely the bottom line. We must aim for a society that is not only sustainable, but that also positively promotes health and well-being in all sections of the human population and in the ecosystems of the biosphere (Figure 7.1). Biosensitivity is a broader and richer concept than sustainability.

Biosensitivity will be a fundamental guiding principle in all spheres of human activity. It will mean biosensitive cities, farms, industries, transport systems, economies, governments and lifestyles.

Box 7.1 Ecologically significant watersheds in cultural evolution

Cultural watershed and approximate starting date	Followed by:
Use and control of fire 200,000 years ago?	Ecological Phase 1 Hunter–Gatherer Phase
Farming 12,000 years ago	Ecological Phase 2 Early Farming Phase
Urbanisation 8,000 years ago	Ecological Phase 3 Early Urban Phase
'Enlightenment' and Industrial Revolution 250 years ago	Ecological Phase 4 Exponential Phase (or 'Anthropocene') Ecologically unsustainable, leading to the collapse of civilisation, with great loss of life — unless humankind moves to Ecological Phase 5.
Biorenaissance Very soon	Ecological Phase 5 Biosensitive Phase Based on understanding the human place in nature. In tune with, sensitive to, and respectful of the processes of life.

Source: Stephen Boyden

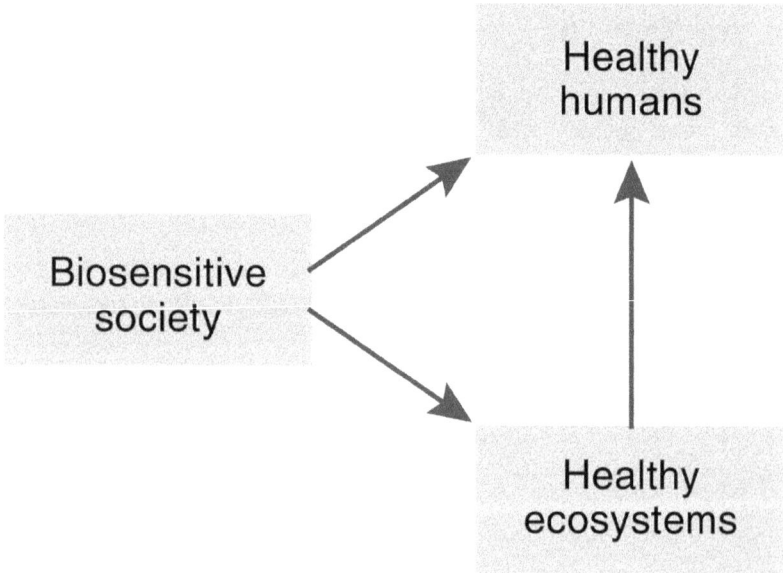

Figure 7.1 The biosensitivity triangle
Source: Stephen Boyden

Human activities and societal arrangements

Promoting health and well-being in all sections of the human population will mean that prevailing conditions will have to be in tune with human biology — that is, they must satisfy the biologically determined physical and psychosocial health needs of our species (see Box 3.2 in Chapter 3). Similarly, social conditions must satisfy the full spectrum of health needs of ecosystems — locally, regionally and globally (see Box 2.1 in Chapter 2).

Some of the most important physical features of a biosensitive society necessary for the attainment of these goals are listed in Box 7.1. This list serves to remind us that the long-term survival of civilisation will require radical changes in many different kinds of human activity.

It will be noted that the last item on this list reads, 'The fifth ecological phase of human history will be free of weapons of mass destruction'. The existence of these weapons is as grave a threat to the survival of civilisation and well-being of humanity as are all the ecological maladaptations discussed in Chapter 5.

The achievement of the necessary physical conditions for biosensitivity will require big changes in societal arrangements. At present, government policies, the economic system and the institutional structure of society are all geared to ever-increasing consumption of resources and, consequently, ever-increasing impact on the living world around us. They are also resulting in extreme differences in the well-being and material wealth of different sections of the population. These arrangements are simply not consistent with biosensitivity and the survival of civilisation.

One of the roles of government will be to oversee a major reduction in the working hours of the labour force — thus diminishing the intensity of technometabolism at the same time as minimising unemployment. The current typical governmental response to unemployment is to attempt to create new and unnecessary jobs. This is ecologically crazy, because it results in further intensification of technometabolism with increased impact on the biosphere. When there isn't enough work to keep everybody busy for eight hours a day, five days a week, the infinitely more sensible approach is to reduce the time that each person spends working.

Box 7.2 Essential physical characteristics of a biosensitive society

Human activities

- minimal use of fossil fuels
- extensive forestation and reforestation and other measures worldwide to sequester atmospheric carbon
- a high proportion of energy used in society coming from clean sources (i.e. not resulting in emissions of carbon or production of dangerous radioactive by-products)
- stable consumption of materials and energy at a sustainable level
- maximisation of local food production
- maintaining a supply of clean water for human consumption, free of pathogenic organisms or harmful chemicals
- farming practices that protect the biological integrity and health of soils
- keeping natural nutrient cycles intact by returning organic waste to farmland
- effective protection of biodiversity in all regional ecosystems and in the oceans
- no release into the atmosphere, waterways or soil of pollutants that interfere with the health of humans and other forms of life — directly (e.g. PM2.5 and SO_2 in the atmosphere, POPs in the soil) or indirectly (e.g. CFCs in the atmosphere)
- people's lifestyles will be:
  - consistent with the biological health needs of the human species (e.g. clean air and water, healthy diet, plenty of physical exercise, the experience of conviviality)
  - consistent with the health needs of the living environment. Emphasis will be on such activities as growing food, making music, dancing, art, theatre, sport, convivial social interaction.

Human population

- a healthy human population with no gross disparities in health and well-being in different sections of the population
- eventual adjustment of global and regional populations to levels that do not cause progressive damage to the planet's ecosystems.

Human artefacts

The built environment will be designed to:

- minimise use of fossil fuels and water and maximise use of clean energy sources
- minimise pollution
- encourage health-promoting activities (e.g. walking, cycling, convivial social interaction)
- maximise biodiversity and opportunities for local food production.

The fifth ecological phase of human history will be free of weapons of mass destruction.

Source: Stephen Boyden

Governments will also be involved in restructuring the workforce and in the transfer of workers in occupations that have undesirable impacts on the environment to jobs that are consistent with ecological sustainability.

The most significant change of all will be in economic arrangements. In a biosensitive society, the ideal of economic growth will be replaced with the ideal of economic health. This will mean that the economic system will:

- be based on economic theory that reflects a sound understanding of the processes of life on which we depend and of the biological limits to human activities on Earth
- ensure the satisfaction of the health needs of all sections of the human community and of the ecosystems of the biosphere
- not result in a continuously increasing rate of use of material resources and energy
- progressively reduce present disparities in material wealth, health and well-being across human populations.

In the educational arena, a core theme at primary and secondary levels will be the story of life and the human place in nature and its relevance to human affairs and everyday life. Students will be alerted to the brainwashing power of culture, and to the need to be constantly vigilant to ensure that the prevailing culture is free from delusions that cause unnecessary human distress or damage to the living environment.

Biosensitive cultures

These essential requirements for the long-term survival of civilisation will not be achieved unless there come about revolutionary changes in the prevailing cultures across the world. In biosensitive societies, these cultures will be characterised by profound respect for the processes of life which gave rise to us, of which we are a part and on which we are totally dependent for our existence. Unlike today, the goal of being sensitive to and in tune with these processes will be seen as of supreme importance. It will be given highest priority on the political and social agenda.

This radical shift in priorities will depend on a wave of new understanding sweeping across the cultures of the world — understanding of the story of life and the human place in nature. Understanding, that is, of the bionarrative. This new understanding will be the pivotal factor in the transition to biosensitivity. All the necessary changes in human activities, such as energy use and consumer behaviour, and in societal arrangements, such as the economic system and government regulations, will follow naturally from this seminal cultural transformation.

The transition

During the past half century there have been many signs of growing awareness among some sections of the community that our present society is heading for ecological collapse. At the international level there has been a series of major conferences on this theme organised by the United Nations, from the Conference on the Human Environment in Stockholm in 1972 to the Conference on Sustainable Development in Rio de Janeiro in 2012. There have also been many important international conferences on specific ecological issues, such as anthropogenic climate change, loss of biodiversity and land degradation.

Over this time, numerous books have been published drawing attention to the fact that the survival of civilisation will require big changes in patterns of human activities on Earth. Early examples from the 1970s include works by Donella and Dennis Meadows, René Dubos and Barbara Ward, Paul Erhlich and Barry Commoner. Since that time there has been an explosive growth of literature on environmental history and environmental philosophy.

Many individuals and groups have come up with ideas for an ecologically sustainable society of the future. In 1972, Edward Goldsmith and others published *A blueprint for survival* in which they argued for a shift to a new kind of society to prevent 'the breakdown of society and the irreversible disruption of the life support systems on this planet'. Today there are many community organisations and NGOs campaigning for a transition to an ecologically sustainable society, such as the transition town movement, the Great Transition Initiative and Inspiring Transition to a Life-sustaining Future.

There are also countless groups focusing on specific ecological issues. To mention but a few local examples in Australia: The Climate Institute, Sustainable Population Australia, SEE Change groups, The Wilderness Society, permaculture groups, Healthy Soils Australia, 350 Australia and Landcare groups.

The emergence of the Greens as a political entity is another indication of a growing concern about the ecological predicament — although election results suggest that this concern is shared by only a small percentage of the electorate.

Despite these encouraging signs, the warnings have not penetrated to the core of the prevailing cultures of the world. We have only to listen to the pre-election speeches of our political leaders for proof of this statement. Although some important measures have been taken here and there to protect aspects of the natural environment, they have not been allowed to interfere with the inexorable thrusts of ever-moreism and market forces. The juggernaut rolls on.

So, while the process of cultural reform is certainly underway, it has a long way to go, and the inevitable counter-reform backlash is very much in evidence. The ecologically maladaptive assumptions of the dominant cultures remain firmly entrenched, and the reform process is clearly in need of a boost.

In our view, the missing ingredient in this reform movement is a concerted effort to promote understanding of the story of life and the human place in nature. As emphasised throughout this book, this bionarrative has great meaning for every one of us and for society as a whole.

All major religions have their stories. This reform movement differs from religious movements in that its story is strictly about the natural rather than the supernatural, and by the fact that the story comes from direct observation of the real world, rather than from the imaginations of mystics and prophets.

Shared understanding of the bionarrative across the cultures of the world is, we believe, a precondition for the survival of civilisation. Only then will the health of the living systems on which we depend be given the highest priority in human affairs. There will be no significant change until this happens.

Once this crucial cultural transformation has taken place there will be wide-ranging changes in the intensity and nature of human activities and a major scaling down of resource and energy use in the affluent nations. Immense effort will be directed to countering the current anthropogenic threats to humanity and the living systems of our planet.

First and foremost, governments and the private sector will treat climate change as a matter of extreme urgency. Strong measures will be introduced to bring about a drastic reduction in the use of fossil fuels, to increase the use of clean energy, to sequester excess carbon in the atmosphere, and to bring an end to coal mining.

Very high priority will also be given to:

- introducing a new economic system that satisfies the health needs of all sections of the human population without resulting in ever-increasing consumption of natural resources
- bringing an end to population growth
- eliminating the current gross disparities in human health and quality of life in different sections of the human population
- protecting biodiversity on land and in the oceans
- protecting the biological integrity of soils and returning nutrients in organic waste to farmland
- increasing local food production
- minimising the release of pollutants that result in damage to living systems
- eliminating weapons of mass destruction.

All this will require enlightened and strong government action, supported by an informed and concerned public.

Action

The most critical need at the present time is, therefore, to set in motion a radical social movement that has the primary objective of awakening the dominant cultures of the world to biological and ecological realities — through spreading understanding of the human place in nature and promoting a vision of society that is truly in tune with, and respectful of, the processes of life.

In most Western countries, the infrastructure to set the ball rolling is already in place in the form of countless NGOs that are working towards ecological sustainability. Unfortunately, they represent only a small section of the community and, so far, their overall effect has been minimal. If all the members of these groups were to devote some of their time and effort to promoting biounderstanding across the community, changes could come about in the prevailing culture in quite a short time.

These NGOs could also join forces to put pressure on UN agencies to become actively involved in the movement. These agencies have the means and the obligation to play a key role in the campaign.

Whether civilisation survives the next hundred years will depend on whether the world's cultural systems come to embrace biounderstanding and take necessary action in time to avert ecological disaster on a massive scale.

Comment

This major shift in understanding is, of course, just the first phase in the reform process. Once the cultural transformation has taken place there will be widespread and informed dialogue at many levels in society on the ways and means of achieving the transition to biosensitivity, and on the new societal arrangements that will be necessary to attain the necessary changes in human activities. This second phase will entail consideration of various options, including legislation, financial incentives, changes in urban planning and transport arrangements — as well, of course, as a major restructuring of the economic system.

This crucial aspect of the reform process is beyond the scope of this book. Our emphasis is on the pivotal first phase — promoting biounderstanding across the community. In fact, the most essential feature of the second 'how' phase will be the requirement that the bionarrative be firmly entrenched in the mindsets of all those involved in the dialogue, whatever their area of specialisation.

The third phase in the transition will be implementation of the new societal arrangements.

Epilogue. Musings

Pessimism and optimism

In previous pages, we have noted that the prevailing cultures across the world are generating human activities on a scale and of a kind that threaten the integrity of the living systems of the biosphere on which we depend. If these activities continue unabated, the ecological collapse of society is inevitable. There are also other highly unsatisfactory features of current society, including the gross disparities in health and conditions of life across different socioeconomic groups and the existence of thousands of weapons of mass destruction.

I have heard it argued that there is no hope of achieving ecological sustainability until we really understand the whole complex biophysical and social system in its entirety, and that much greater effort should be aimed at achieving such understanding through systems modeling. In my view, despite recent advances in systems theory and information technology, the complexity of the system is such that this understanding will always be beyond us.

However, all is not lost, because I suggest we don't need to understand all the intricacies of this massive and extremely complicated system in order to move forward to a biosensitive society.

The approach that we advocate is indeed much simpler. All that is required initially is a single, if highly significant, change in the system. This change will involve all the world's prevailing cultures coming to embrace a sound understanding of the bionarrative. Such biounderstanding at the heart of these cultures will have far-reaching repercussions throughout the whole of society — the ripple, or butterfly, effect.

As an outcome of this new understanding, these cultures will share profound respect for the processes of life in and around us, and they will give highest priority in human affairs to the attainment of harmony with these processes. This life-oriented worldview and set of priorities, or *bioperspective*, will be the pivotal factor in the ecological survival of civilisation. All the necessary changes in human activities (e.g. energy use, consumer behaviour, forestry, lifestyles) and in societal arrangements (e.g. the economic system, government regulations, population policies) will follow naturally from this seminal cultural transformation. Shared biounderstanding across all societies is thus a precondition for the future well-being of humankind and the rest of the living world.

We can call this transformation a biorenaissance, because in the distant past hunter–gatherer and early farming cultures understood that humans are part of nature and completely dependent on other forms of life for their well-being and survival, and they held deep respect for the living world.

What is the likelihood of this cultural enlightenment coming about soon enough to avert ecological disaster on a massive scale? I am rather pessimistic. The maladaptive assumptions of the prevailing cultures are deeply ingrained. The notion that economic growth must take precedence over all other considerations and general ignorance of biological and ecological realities do not auger well for the future.

On the other hand, although I think it unlikely that effective cultural reform will come about soon enough to avert catastrophe, I am optimistic enough to think it is not impossible. So long as this is the case, I feel strongly that those of us who understand the nature and seriousness of the situation should continue to do all we possibly can to hasten this critical revolution in understanding and thinking.

One thing that could accelerate the cultural transformation would be for the ideology behind the movement to be given a name. All major political ideologies and religious belief systems have names — Marxism, fascism, socialism, capitalism, Buddhism, Islam, Christianity, Zionism and so on — and yet this life-oriented ideology, ancient as it is, has yet to be bestowed with a universally accepted name.

The term 'environmentalism' has been in common use since the 1960s, but it is not adequate for our purpose. The cultural and societal transformation that will be necessary to make the transition to biosensitivity will need to go much deeper than mere concern for 'the environment'. It needs to be a whole new mindset that understands, celebrates and respects life on Earth and that places the health and well-being both of humans and of the ecosystems of the biosphere right at the top of the social and political agenda.

The adjective 'green' is also widely used today, meaning 'concerned with or supporting protection of the environment as a political principle' (*Oxford Dictionary*), and 'greenism' has been used for the underlying philosophy. While greenism is an all-important aspect of the ideology that we are talking about, it is less comprehensive, and I find it an awkward term.

Some years ago I coined the word 'biorealism' for this purpose — but again I felt it was not adequate.[1] Since then I have wasted many hours trying to find a more appropriate term. Possibilities that have come to mind include biocentrism, biosocialism, biophilism, bioradicalism, bioempathism, bioactivism, biofuturism and life-ism. But none of these words really fits the bill, and anyway it turns out that all of them are already in use, and none is defined in terms that correspond to our meaning.

So we must patiently wait for someone to come up with an appropriate name for the ideology on which the survival of civilisation will depend.

My personal vision of the cultural reform process is summarised in Figure 8.1. Some readers will see these ideas as hopelessly naive and unrealistic. Pie in the sky stuff. Perhaps they are; but if so, then I believe there is little hope for humanity.

---

1   S. Boyden, 2004. *The biology of civilisation: Understanding human culture as a force in nature.* UNSW Press, Sydney, p.176.

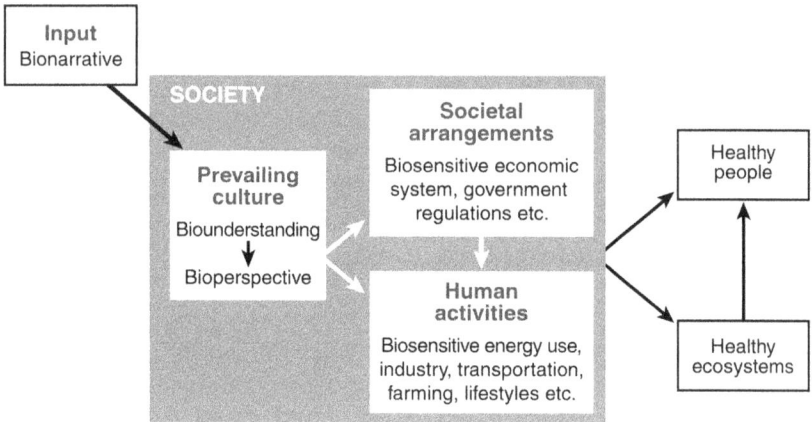

Figure 8.1 The biotransition

Source: Stephen Boyden

Tenets

- The massive growth of the human population and the great intensification of resources and energy use by human society are unsustainable ecologically. If present trends continue unabated the collapse of civilisation is inevitable.

- There are, also, other highly unsatisfactory features of the human situation today, like the major disparities in health and conditions of life across different socioeconomic groups and the existence of thousands of weapons of mass destruction.

- The best hope for the survival of civilisation lies in a speedy transition to a society that is truly sensitive to, in tune with, and respectful of the processes of life. This is a biosensitive society. It will be a society that promotes health and well-being in all sections of the human population and in the ecosystems of the biosphere. This will require big changes in the scale and nature of human activities on Earth.

- By far the most essential difference between a biosensitive society and that in which we live today will lie in the world view and priorities of the prevailing culture. In a biosensitive society, this culture will embrace a sound understanding of the story of life on Earth and the human place in nature. This is biounderstanding.

- As an outcome of this shared biounderstanding, the prevailing culture will hold profound respect for the processes of life. Unlike today, the goal of being sensitive to, and in tune with, these processes will be at the top of the political and social agenda.

- This fundamental shift in priorities, in what matters most, will be the key factor in the survival of civilisation. All the necessary changes in human activities (e.g. energy use) and in societal arrangements (e.g. the economic system) will follow naturally from this seminal cultural transformation.

- Shared biounderstanding across all parts of society is thus a precondition for the survival of civilisation.

An ethical digression

> The richness I achieve comes from nature, the source of my inspiration. I have no other wish than to mingle more closely with nature and I aspire to no other destiny than to work and live in harmony with her laws.
>
> — Claude Monet[2]

> Nature is my god. To me, nature is sacred. Trees are my temples and forests are my cathedrals.
>
> — Mikhael Gorbachev[3]

For some of us, the bionarrative has meaning beyond the purely practical. It has ethical significance. This is not to say that this story tells us what is good or evil. The bionarrative comes from the sciences, and the sciences tell us nothing about morality. They do not tell us whether it is right or wrong to cause needless pain or distress in humans or other animals, or whether it matters if the human species comes to an early end through its own activities and arrogance.

Another weakness of the bionarrative is that, while it tells us more and more about the biophysical world, and what happened in the past and how things work today, it does not explain everything. It does not explain existence as such. It suggests that the universe began with a big bang, but does not explain why there was a big bang, or what existed before the big bang, or whether there was a before.

It has been said that life itself is ultimately explainable in terms of the laws of physics. These laws will lead, willy-nilly, to life on suitable planets in the universe. There is, however, a difficulty with this proposition. Atheists are confronted with the same problem as those who believe in God. The latter are faced with the conundrum 'Who created God?' Atheists must ask: 'Where did the laws of physics come from?'

---

2   www.wisdomquotes.com/quote/claude-monet.html.
3   'Nature is my god', interview with Fred Matser in *Resurgence*, no. 184, Sep.–Oct. 1997: 14–15.

And, indeed, what a wondrous set of laws — laws that have led to the eventual coming into being, starting from the primordial mass of matter and energy, of the myriads of living organisms that make up our biosphere, as well as the plays of Shakespeare and the symphonies of Beethoven.

On reading some recent discussions on why humans develop religious belief systems, there is one thing that strikes me as odd. I am puzzled by the fact that they say little about the 'thank you factor'.[4] The following quotation from Ian Warden, a columnist in the *Canberra Times*, puts it nicely:

> True believers are so very lucky. So much in life is so very lovely (a morning of whales! of albatrosses! of literature! of wearing fancy dress!) that, though a non-believer now, it still seems spiritually impolite to know such happiness, but to not be able to say a heart-felt 'thankyou' for it.[5]

I share Warden's predicament. The processes of life have brought me into being. They have given me my life, my body, my senses, and my capacity to think, to love and to enjoy. They have given me the opportunity to look around, to marvel, to feel and to choose pathways of action. I, too, would like to say a heartfelt 'thank you' — for this wondrous gift.[6]

My own philosophy of life has much in common with deep ecology.[7] One definition of deep ecology reads as follows:

---

4   R. Winston, 2006. *The story of God: A personal journey into the world of science and religion.* Bantam Books, London; and, R. Dawkins, 2006. *The God delusion.* Bantam Press, London.

5   *Canberra Times*, 4 Nov. 2006, p. 2.

6   I appreciate that there are many people who would not share this thankfulness. Cultural evolution has brought about a situation such that many hundreds of millions of humans live very miserable lives in conditions of extreme squalor and deprivation. It seems that even Buddha felt no sense of thankfulness. His aim was to escape from the shackles of ordinary life on Earth. I am among those who are more fortunate.

7   This movement owes its origin to the writings of Arne Naess — see, for example, 'The shallow and the deep, long-range ecology movements, *Inquiry*, 16, 1973: 95–100. It is a pity about the name 'deep ecology'. According to my dictionary 'ecology' means *the study of the relationships between living organisms and their environment*. 'Deep ecology' would therefore mean *the deep study of the relationships between living organisms and their environment*. This, of course, is not at all what deep ecologists mean by the term.

Deep ecology is an approach to ethics that holds that the non-human environment has intrinsic value that is independent of human interests. Deep ecology is a reaction to anthropocentric approaches to the environment which hold that the environment has value only as a means of promoting human interests. Deep ecologists view the value of human activities in a larger environmental context.[8]

Certainly, I find myself in general agreement with 'The Deep Ecology Platform', one version of which reads:

1. All life has value in itself, independent of its usefulness to humans.
2. Richness and diversity contribute to life's well-being and have value in themselves.
3. Humans have no right to reduce this richness and diversity except to satisfy vital needs in a responsible way.
4. The impact of humans in the world is excessive and rapidly getting worse.
5. Human lifestyles and population are key elements of this impact.
6. The diversity of life, including cultures, can flourish only with reduced human impact.
7. Basic ideological political, economic and technological structures must therefore change.
8. Those who accept the foregoing points have an obligation to participate in implementing the necessary changes and to do so peacefully and democratically.[9]

Some deep ecologists put a great deal of emphasis on what they call 'deep experience', which is described as a semi-religious experience.

Deep ecologists disapprove of what they call 'anthropocentrism'. If by anthropocentrism they mean an attitude that sees humans as separate and superior to nature, then I am in entire agreement. Such an attitude is based on ignorance and can lead us into big trouble. But, if anthropocentrism is taken to mean the tendency of humans to be especially interested in themselves and their own well-being, then I think we are stuck with it. It is a natural human characteristic,

8   See www.scicom.lth.se/fmet/ethics_03.html.
9   See Stephan Harding, 'What is deep ecology', www.schumachercollege.org.uk/learning-resources/what-is-deep-ecology.

the product of our evolutionary background. In my view, however, this kind of anthropocentrism is consistent with the protection and enjoyment of the rest of the living world.

Deep ecologists and like-minded people are in a minority in our society. The ideal of 'conquering nature' is still with us today. It goes back over 300 years to the early days of the so-called Enlightenment. For example, Descartes wrote of the purpose of science as part of the struggle to 'render ourselves the masters and possessors of nature'.[10]

The notion of conquering nature was also part of the central dogma of Mao Zedong's Cultural Revolution, which was contrary to earlier, established Chinese philosophy.[11]

A few years ago, I came across the depressing online journal *Capitalism Magazine*. The titles of a couple of articles on the environment were as follows: 'Reject environmentalism, not DDT'; and, 'Wasting billions on the green agenda'. There was also an article called 'Industrialisation and the environment' in which the author describes what he calls the 'nature-as-sacred creed'. He finishes his article with the following words:

> My conclusion [is that] we should reject this anti-human creed [the nature-as-sacred creed] and uphold man's right to achieve his full glory by using his rational mind to conquer nature for the purpose of enhancing and enjoying life.

My position

Having apparently dismissed science as a source of ethical judgement, I will now describe how, for me, the sciences and, in particular, the life sciences and biohistory, have had a profound effect on my world view and values.

---

10   Rene Descartes, *Discourse on Method*, 1637.
11   See Judith Shapiro, 2001. *Man's war against nature: Politics and the environment in revolutionary China*. Cambridge University Press.

First, I must say that I am not attracted to the idea of a supernatural being or god who has created each species of animal and plant separately and who is monitoring my behaviour and who will, if I believe in him or her, take me up to Heaven when I die.[12]

All religions have their stories. Christians, Muslims, Buddhists and Hindus all have their stories. I have my story. It is the story of life on this planet and of the evolutionary emergence of humankind and of the interactions between humans and the rest of the living world. It is a story based on people's conscious observations and powers of reason. It is not the stuff of dreams. It comes to us through the natural sciences. It is a truly amazing story and, in my view, it is likely to be much nearer the truth than any of the stories that are products of the imaginations and revelations of prophets and soothsayers.

The more I learn about the story of life on Earth and of the details of living processes — through personal observations and through the scientific observations of others — the greater is my sense of wonder. And, the more I learn, the greater my sense of respect for life and the creative forces that gave rise to the living world and to me.

I am thinking not only of the natural environment, with all its diversity and beauty, but also of the incredible and extremely complicated set of processes that go on inside my own body and that have kept it going for over 91 years.

For me, the bionarrative is a great source of inspiration. It is a driving force in my life, but it still leaves an unexplained mystery. This underlying mystery is ultimately responsible for me, and for all of life on Earth. It is unscientific to deny the existence of this mystery, just as it is rather silly to dream up an explanation in the form of a supernatural human-like being. I am content to accept the mystery as unexplainable at the present time, and I suspect that it will always remain so.

---

12   And yet I rather enjoy some of the ritual associated with religion. Of the various kinds of belief systems that humans have concocted over the millennia, I have most sympathy with those that involve nature worship, which seems to me to make much more sense than the worship of any one particular god — because nature is something we can all experience directly, something we can all see. We are part of it and without it we could not exist. And for me it is wondrous and beautiful.

Sometimes my appreciation of the mystery, along with the experience of natural beauty, gives rise to feelings that I suppose could be described as spiritual — an apparent awareness of another dimension to reality, associated with a deep feeling of awe and reverence. Whatever the chemical phenomena in my brain cells that might lie behind such feelings, they are a vitally important part of my life experience.

My scientific understanding of the story of life, together with the feeling of deep thankfulness, generate in me a very strong wish to live in harmony with, and to be supportive of, the processes of life, and to be protective of nature when it is threatened by the actions of humankind. They lead me to believe it is wrong to kill living organisms needlessly, to cause unnecessary pain or distress in humans or other animals, or to trash the living systems that brought us into being and that underpin our existence. I feel it is right to seek to live in tune with the processes of life, and I see the promotion of health and well-being in fellow humans and in the living systems of which we are a part as what matters most.

We humans have all received the gift of life — and I see it as very wrong to deliberately deprive any other human of the opportunity to enjoy being alive, just as it is right to help others overcome threats to their existence and well-being.

These are ethical statements, and they are an outcome of my scientific understanding of the story of life on Earth and the human place in nature. Although the underlying source of these tenets is science, science does not, however, prescribe them. They come from within me.

I am among those who share Albert Schweitzer's deep sense of reverence for life.[13]

<hr>

13 Albert Schweitzer lived from 1875 to 1965. His Reverence for Life philosophy is well described in en.wikipedia.org/wiki/Reverence_for_Life.

Glossary

Adaptation
: A response of an individual (or a population) to an environmental threat that renders the individual (or population) better able to cope with the threat.

Artefacts
: Components of the environment that are manufactured by humans.

Arthropod
: An invertebrate animal with jointed legs, such as an insect, spider or crustacean.

Basic behaviour
: Behaviour shared by all humans, such as eating when hungry and seeking the approval of one's in-group. The specific manifestations of basic behaviour can vary according to circumstances and cultural influence (see specific behaviour).

Biohistory
: The study of the history of life on Earth up to the present moment, including the interactions between humans and the living systems of the biosphere.

Biological advantage
: Favouring health, survival and likelihood of successful reproduction.

Biometabolism
: The inputs, internal uses and outputs of materials and energy involved in biological processes within living organisms.

Bionarrative
: The story of life on Earth, including the emergence and growth of human civilisation and the interactions between humans and the rest of the living world.

Biorenaissance	A reawakening of the world's prevailing cultures to the reality that humans are living beings, products and part of nature and totally dependent on the processes of life for their wellbeing and survival. Keeping these processes healthy is top priority because everything else depends on them.
Biosensitive society	A society that is sensitive to, in tune with, and respectful of the processes of life and that promotes health and well-being in all sections of the human population and in the ecosystems of the living environment.
Biosphere	All the living organisms on the planet and the physical environment with which they interact.
Biota	Living organisms.
Civilisation	All human societies with economies based on farming (i.e. Ecological Phase 2, 3 and 4 societies).
Counter-reformers	People who actively oppose cultural reform.
Cultural delusion	A mistaken cultural assumption.
Cultural maladaptation	A cultural delusion that leads to behaviour that causes unnecessary distress in humans or that causes unnecessary damage to other living systems in the biosphere.
Cultural reform	Cultural responses aimed at overcoming undesirable consequences of cultural maladaptations.
Cultural system	The abstract aspect of human situations, which includes culture (language, beliefs, assumptions, values, etc.) and cultural arrangements (e.g. legislation, economic arrangements, institutional structure, etc.).

Culture	The shared and accumulated knowledge, beliefs, assumptions, values and technical knowledge that are passed from one individual to another, from one group to another and from generation to generation, mainly through the use of learned symbols, as in speech and writing.
DNA	Deoxyribonucleic acid, the essential self-replicating component of the genetic apparatus of living cells, responsible for the transmission of hereditary characteristics from parents to offspring.
Domestic transition	The transition in human history from the Hunter–Gatherer Phase 1 to the Early Farming Phase 2.
Ecological phases	Phase 1 — the Hunter–Gatherer Phase Phase 2 — the Early Farming Phase Phase 3 — the Early Urban Phase Phase 4 — the Exponential Phase
Ecologically sustainable	A population (or society) is ecologically sustainable when the ecosystems on which it depends — local, regional and global — maintain their capacity to satisfy the health and survival needs of that population (or society).
Ecologically unsustainable	A population (or society) is ecologically unsustainable when the ecosystems on which it depends are progressively losing their capacity to satisfy the health and survival needs of that population (or society). Such a situation can come about either because the ecosystems are degrading and/or because the population is expanding beyond sustainable levels in terms of supply of food and water.

Evolutionary health principle	The principle that if an animal is removed from its natural environment, or if its environment changes in some significant way, then it is likely that the animal will be less well adapted to the new conditions, and will consequently show some signs of physiological or behavioural maladjustment.
Extrasomatic energy	Energy used by humans outside the human body in various technologies. It is thus distinct from *somatic energy*, which is the energy used in metabolic processes within the human body and provided by food.
Genetic engineering	The deliberate scientific modification of the characteristics of an organism by manipulating its genetic material, sometimes involving the incorporation of genetic material from other organisms.
Greenhouse effect	The warming of the Earth's surface and atmosphere due to the presence of certain gases in the atmosphere (e.g. water vapour, carbon dioxide) that capture heat radiating from the Earth's surface and reradiate it downwards towards the Earth's surface.
Hominid	Any primate belonging to the family *Hominidae*, which includes modern humankind as well as earlier upright-walking species and the great apes.
Human activities	Human behaviour at the societal level (e.g. manufacturing, farming, constructing buildings, making war).
Industrial transition	The transition from Early Urban Society (Ecological Phase 3) to modern Exponential Society (Ecological Phase 4).
Lignin	A complex molecule occurring in certain plant cell walls, making the plant rigid.
Meliors	Experiences that promote a sense of enjoyment and well-being.

Natural environment of an animal or plant	The natural environment of an animal is the environment in which it evolved, and to which it is therefore genetically adapted through natural selection. For humans, the natural environment is that of the hunter–gatherer, which was the only environment known to our species for many thousands of generations.
Photosynthesis	The process by which energy in the form of sunlight is captured in the leaves of green plants and converted into chemical form through the action of chlorophyll.
Primeval	Referring to the first hunter–gatherer phase of human existence.
Selective advantage	Promoting survival and successful reproduction.
Shared knowledge	Knowledge shared by the majority of the members of a society.
Societal arrangements	The legislative, economic and institutional arrangements of society.
Somatic energy	The energy used in metabolic processes within living organisms.
Specific behaviour	Specific behavioural manifestations of basic behavioural tendencies.
Stressors	Experiences that tend to promote a state of stress or distress.
Technoaddiction	The state of having become dependent on a particular technology for normal living.
Technometabolism	The inputs, uses and outputs of energy and materials of a society resulting from technological processes and taking place outside human bodies.
Universal health needs	The biologically determined health needs of the human species.

Previous books by the author

as editor, 1970. *The impact of civilisation on the biology of man.* Australian National University Press, Canberra.

1979. *An integrative ecological approach to the study of human settlements.* MAB Technical Note No. 12. UNESCO, Paris.

with S. Millar, K. Newcombe and B. O'Neill, 1981. *The ecology of a city and its people: The case of Hong Kong.* Australian National University Press, Canberra.

1987. *Western civilization in biological perspective: Patterns in biohistory.* Oxford University Press, Oxford.

with S. Dovers and M. Shirlow, 1990. *Our biosphere under threat: Ecological realities and Australia's opportunities.* Oxford University Press, Melbourne.

1992. *Biohistory: The interplay between human society and the biosphere — past and present.* Unesco, Paris.

2004. *The biology of civilisation: Understanding human culture as a force in nature.* UNSW Press, Sydney.

2005. *People and nature: The big picture.* Nature and Society Forum, Canberra.

2011. *Our place in nature: Past, present and future.* Nature and Society Forum, Canberra.

Index

www.ingramcontent.com/pod-product-compliance
Lightning Source LLC
Chambersburg PA
CBHW040149270326
41927CB00029B/3428